The Psychology of Child Firesetting

DETECTION AND INTERVENTION

By

Jessica Gaynor, Ph.D.

and

Chris Hatcher, Ph.D.

BRUNNER/MAZEL *Publishers* • New York

Library of Congress Cataloging-in-Publication Data

Gaynor, Jessica, 1949–
The psychology of child firesetting.

Bibliography: p. 209
Includes index.
1. Pyromania in children. 2. Fire behavior in children. 3. Arson. 4. Child psychotherapy. I. Hatcher, Chris, 1946– II. Title. [DNLM: 1. Impulse Control Disorders—in infancy & childhood. 2. Impulse Control Disorders—rehabilitation. WM 176 G287p]
RJ506.P95G38 1987 618.92′85843 86-21642
ISBN 0-87630-445-5

Published by
BRUNNER/MAZEL, INC.
19 Union Square
New York, New York 10003

MANUFACTURED IN THE UNITED STATES OF AMERICA

Contents

SECTION II. INTERVENTION

Preface

The work for this book began nearly a decade ago when Chris Hatcher brought together Pamela McLaughlin and the senior author to discuss an idea Pam had about how to help youngsters showing patterns of recurrent firesetting behavior. The idea was to create partnerships between firefighter volunteers and young boys involved in firesetting. This idea was developed in response to recent statistical reports indicating that over half of those arrested for the crime of arson were young boys under 18 years of age. The United States Fire Administration awarded a research and development grant to Fire Chief Andrew Casper's San Francisco Fire Department to implement this program and test its feasibility and impact. The senior author's expertise in program planning and evaluation was applied to assess the effectiveness of pairing firefighter volunteers with youngsters in reducing the incidence of fireplay, firesetting, and juvenile-related arson, and improving the quality of life for these children and their families.

At this point the research focused on investigating what was ''known'' about the relationship between youngsters and fire. The senior author learned about the pioneering studies of Lewis and Yarnell describing pathological firesetting youngsters and the subsequent clinical and empirical literature stimulated by this original work. Although the 20-year-period between 1950 and 1970 generated several important studies describing the personality characteristics of youthful firesetters, there were few attempts to develop effective intervention strategies to eliminate recurrent firesetting behavior and the accompanying psychopathology. However, during the 1970s two different and significant trends emerged in the treatment of pathological firesetting. First, there was a shift in the literature from studies focused on describing the personality characteristics of youthful firesetters to research suggesting the success of various psychotherapies in eliminating firesetting behavior. Case studies reported the effective application of outpatient and inpatient treatment programs, including psychodynamic, cognitive-emotion, behavior, and family psychotherapy methods to stop firestarting behavior and remediate the accompanying psychopathology. Second, a coalition of professionals, including Captain Joe Day of the Los Angeles County Fire Department and psychologist Kenneth Fineman, Ph.D., began to develop community-based intervention strategies to help youngsters involved in firesetting. Supported by the United States Fire Administration, this group developed the first screening procedure to be utilized by firefighter personnel to identify and evaluate youngsters "at risk" for becoming involved in fireplay, firesetting, and juvenile-related arson. Hence, these two significant advances represented important movement in the evaluation and treatment of child firesetting.

The intervention program initiated in the San Francisco Fire Department combined the application of the screening and evaluation procedure developed by Captain Day and Dr. Fineman with Pamela McLaughlin's idea of helping firesetting youngsters by pairing them with firefighter volunteers. In addition, the San Francisco program included the availability of mental health consultants to conduct psychological assessments of children and their families as well as to provide training for firefighter volunteers on how to work with firesetting youngsters. The research and development phase indicated that this intervention method was effective in eliminating firesetting behavior and improving the quality of life for participating children and their families. As a result of this work, the

National Firehawk Foundation was created to promote and distribute the Firehawk Children's Program, which represents one example of a community intervention strategy designed to identify and treat youthful firesetters.

The material in this book is the product of applying a number of different methods to learn about the psychology of child firesetting. First, there has been a comprehensive review of the available research describing the characteristics of youthful firesetters, their behavior, and the various intervention methods designed to help them. Second, both authors have clinically evaluated and treated several cases of firesetting children and their families. Third, the authors have participated in the planning and evaluation of community-based intervention programs for youthful firesetters. Finally, the experience of a network of professionals, representing a wide variety of fields, including the fire service, mental health, medicine, social work and juvenile justice, has been compiled and communicated throughout the chapters of this book. Research, clinical experience, program planning and evaluation, and networking with a diverse group of professionals are the four primary data collection methods utilized in this effort to present the most exhaustive and current information on the evaluation and treatment of youthful firesetters.

The authors wish to dedicate this book to all those family members, friends, colleagues, and organizations who are helping to make a meaningful contribution to fire safety, prevention, and the mental health of our youngsters.

Jessica Gaynor
Chris Hatcher

August 1986

The Psychology of Child Firesetting

DETECTION AND INTERVENTION

Section I
Detection

1

Firesetting—Our Burning Youth

THE HIDDEN MAJORITY

Consider the following stories taken from actual clinical cases.

Johnny is seven years old and lives with his recently remarried mother and stepfather. Johnny's father died in an airplane accident when Johnny was two. There are no other children in Johnny's new family. For Christmas Johnny's stepfather bought him a chemistry set. One afternoon Johnny was playing alone in his room. He was curious to see whether the new chemical solution he developed would ignite. He found some of his mother's matches and tried to light his test tube. The burning match accidentally fell to the floor, and the bedroom curtains caught fire.

Donna is nine years old and lives alone with her mother in a small one-bedroom apartment. She has never known her father. Her mother suffers from several different medical problems, which often confine her to bed, leaving Donna to manage most of the housekeeping. Donna has been setting small fires around the house since shortly before her fourth birthday. She always hides the burned refuse behind kitchen appliances and denies firesetting. One evening after dinner, Donna's mother, complaining of a backache, said she was going to bed. She asked Donna to wash the dishes and then finish her homework. One hour later Donna's mother awoke to the smell of smoke, ran to the kitchen, and found Donna burning hot oil in the frying pan.

Gary is 14 years old and the third child of a two-parent, intact family. In the summer of Gary's thirteenth year he set his first fire with two other adolescent friends. They were playing a game of "dare" in the abandoned field next to his home. One of the boys was successful in using a lighter to ignite dry grass. The following summer, while having a "smoke" in his grandfather's barn, Gary accidentally dropped his cigarette in a box, the box smoldered, and eventually the entire barn caught fire. As a result, Gary was ordered by juvenile court to spend six months in a residential home. Cigarettes, matches, and lighters were freely dispensed in this setting, and within six weeks Gary had set several small fires and one large fire in a school gymnasium. Gary was sent back to juvenile court for another disposition.

These three youngsters and their firesetting experiences represent a small part of a much larger problem confronting today's families and communities. Although the circumstances and severity of these firesetting incidents varied, all of these youngsters were involved in potentially life-threatening situations. Because of the consequences of firesetting, it is time we begin to ask some serious questions about the problem. How many youngsters are actually involved with fire? Is this yet another new social problem facing today's youth? What are the scope and severity of the firesetting problem? It is only recently that we have been able to answer these questions with some sort of statistical accuracy. The answers to these questions provide a shocking, but realistic appraisal of the work that needs to be achieved to remediate the current firesetting problem and prevent the burning of our youth.

The Facts Reveal the Majority

Fire is an essential part of human life. The control and mastery of fire have helped build modern civilization, but its destructive effects have leveled modern cities and have been used as weapons in times of war. The power of fire both attracts and repels. It symbolically stands for love and hate, destruction and recreation. Fire is an important part of our relationship to the physical world. This complex relationship between human behavior and fire begins early in childhood. Fire interest in children is universal.

Research shows that children express a natural curiosity about fire. Studies of randomly sampled "normal" children, ranging from grades kindergarten to 4, showed that a significant proportion (30% to 60%) either expressed an interest in or had actually played with fire once or twice (Block, Block, & Folkman, 1976; Kafry, 1978). In addition, there is evidence suggesting that children's involvement with fire starts as early as the age of three (Folkman, 1966; Block & Block, 1975). In many instances, if this early attraction to fire is handled in a constructive manner, children can learn the necessary behavior to handle fire competently in a supervised setting. Unfortunately, children's interest in fire often goes unrecognized, resulting in actual incidents of fireplay. Data show that if children play with fire once, their fireplay will result in fires 33% of the time. If they play with fire more than once, their fireplay is likely to result in fires 81% of the time (Lewis & Yarnell, 1951). Hence, what begins in youngsters as a natural curiosity about a significant physical event in their environment, fire and firestarting, can end with serious and life-threatening consequences.

Accidental, Intentional, and Criminal

Data show that many of today's youngsters express a healthy curiosity about fire, but the probability remains high that this curiosity will result in accidental fires. Accidental fires represent a significant cause of death and injury to children (Kafry, Block, & Block, 1981). In fact, statistics show that the majority (60%) of child-set fires are accidental and represent first or second incidents of firestarting (FEMA, 1979; McLaughlin, 1983). However, accidental fires do not represent the only way in which youngsters become involved in firesetting. Recent work has shown that child-set fires can be the result of repeated, intentional firesetting behavior. In addition, there is evidence suggesting that youngsters are being arrested and convicted of arson and arson-related crimes. Therefore, youthful involvement in firesetting can be accidental, intentional, or criminal, depending on the motivations and circumstances.

There are several sources of statistical information indicating the frequency with which children have a hand in their own fire accidents. It was reported that 9.7% of the total number of accidental home fires each year are caused by children playing with matches (California Division of Forestry, 1977). In addition, in the United States accidental fires in buildings caused by children playing with matches increased from 6.2% in 1955 to 10.8% in 1973 (Kafry, Block, & Block, 1981), and in the United Kingdom children playing with fire was the second leading cause of fires in dwellings during the early 1970s (Whittington & Wilson, 1980). Two studies looking at the number of children burned in fires showed that the majority of fire accidents and subsequent burn injuries resulted from children playing with fire (touching hot objects, playing with flames, lighters, and matches) (Wilmore & Pruitt, 1972; Gladston, 1972). Hence, not only is the number of accidental fire deaths and injuries a major threat to the lives of children, but the alarming frequency with which youngsters themselves are involved with these accidental fires is a serious problem.

Unfortunately, youthful fire behavior goes beyond involvement in accidental fires. Most accidental fires attributed to youngsters are labeled "children playing with matches." This is a very misleading description of a somewhat more complicated situation. If accidental fires can be associated with those youngsters who, out of curiosity or experimentation, play with fire once or twice, then the "children playing with matches" label is accurate. In fact, it has been reported from a number of different sources that the majority of child-set fires (60%) are ignited by what is referred to as the "curiosity firesetter" (FEMA, 1979; McLaughlin, 1983). However, over a two-year period from 1980 to 1982, it was discovered that a significant number of youthful firesetting incidents (35%) reported to a fire department in a metropolitan area were the result of recurrent child and adolescent fire behavior (McLaughlin, 1983). There is additional evidence suggesting this figure of repeated firesetting incidents reported to a fire department may be an underestimation of the problem. A recently completed study surveying a random sample of normal grammar-school children found that 40% had played with fire "several to many times" and only three or four of the firesetting incidents had been reported to the fire department (Kafry, Block, & Block, 1981). Therefore, these data indicate that although the majority of youngsters become involved in firesetting out of curiosity and probably set one or two small fires accidentally, there also is a significant number of youngsters who are involved in repeated episodes of firesetting. This information is particularly disturbing since it has been clearly demonstrated that the greater the number of youthful firesetting incidents, the higher the probability that they will result in large and dangerous fires (Lewis & Yarnell, 1951).

Not only are youngsters involved in accidental as well as recurrent firesetting, but the most frequently occurring criminal activity for which children are convicted is arson. In 1977, it was reported that children under 13 constitute 17.1% of all arson arrests in this country (FBI, 1977). In 1984, 43% of all people arrested for arson were under 18 years of age, an increase of 7% from the previous year, and 64% were under 25 (FBI, 1985). The statistical picture of youngsters involved in firesetting, from accidental play to criminal activity, is a frightening and life-threatening situation on which families and communities cannot turn their backs. How is it that this relatively high incidence of youthful firesetting has gone undetected until now?

Hidden by Innocence and Fear

There are several reasons why youthful firesetting has only recently been recognized as an important problem searching for immediate solutions. These factors not only obscured the youthful firesetting problem, but they retarded endeavors to break ground in developing effective intervention and prevention strategies.

First, until the late 1970s there was a remarkable lack of uniform statistical reporting on who set fires. Statistics were available from many different sources on fire deaths, dollar damages, and sources of fires, but there were few attempts to compile the data and examine trends over time. When, for the first time, a statistical overview was presented regarding fire incidences and their causes, the results were both informative and shocking. Two issues deserved particular attention. In both the United States and the United Kingdom, there was a significant increase in ''children playing with fire'' as a major cause of fires in buildings (Kafry, Block, & Block, 1981; Whittington & Wilson, 1980). In addition, the number of fires caused by confirmed or suspected arson in the United States rose by 366% from 1964 to 1974 (Teague, 1978). By 1981, 122,610 fires in this country were attributed to arson; the United States had the highest arson rate in the world with one out of every four fires intentionally set, and, perhaps most disturbing of all, over one-third of all convicted arsonists were children under 18 years of age (FBI, 1982). These startling statistics led fire service leaders to declare youthful firesetting a high priority and called for an immediate investigation of the problem (FEMA, 1979).

A second major factor obscuring the problem of youthful firesetting is that parents were hesitant to seek help for their children. This hesitancy was based on several concerns, many of them motivated by fear. Some parents felt that if they disclosed their children's firesetting, there would be serious consequences for their family. Parents feared alienation from other families and friends. They also feared their children's expulsion from school and other social and community activities. In addition, because many children use recurrent firesetting as a cry for help to call attention to other psychological or social conflicts, parents feared they would be forced to disclose additional family problems. The general attitude was to play down the serious nature of firesetting and attempt to ignore it in the hope that it would go away.

Perhaps the most common reason that parents hesitated to seek help was fear of intervention by the authorities. Many police departments and court authorities across the country view firesetting as the first clue to children's future criminal careers. For example, in one major metropolitan city youthful firesetters are photographed and thumbprinted to help build the solid case which, authorities believe, inevitably will follow. In addition, once children have been brought into the juvenile justice system as a result of their firesetting, there is little attempt to find out the nature and cause of their behavior. Rather, the more typical attitude is to incarcerate these youngsters for the purpose of teaching them a lesson about what it is like to be punished and treated as a criminal. Hence, lack of trust and fear of punishment by authorities have prevented parents from seeking the help they need for their youngsters involved in firesetting.

A final factor hiding the importance of the youthful firesetting problem is the lack of scientific information about the causes and treatment of firesetting behavior in children. Although there has been a large number of medical, criminological, and fire publications concerning the psychopathology of firesetting, relatively little is known about the origins and treatment of the behavior. This is in part due to the lack of systematic, well-controlled studies presenting reliable data. Many studies are based on impressions from authors who have worked on individual cases. In addition, data on firesetting come from many sources such as juvenile court, mental health clinics, and other community agencies, which makes it difficult to compare samples and draw generalizable conclusions. As a result, although there are some studies providing adequate descriptive data on children involved in firesetting, there have been few systematic attempts to delineate normative data describing the youthful firesetting population. Without such normative samples, it also becomes difficult to investigate the effectiveness of intervention programs designed to help youthful firesetters.

Help for Identified Firesetters

What about Johnny, Donna, and Gary? What kind of help is available to the youngsters whose brief case histories were presented? Obviously, the three cases differ in terms of the severity of their exhibited firesetting behavior. Each youngster can be looked at first in terms of what typically might happen as the result of the firesetting, and second, what should or ideally might happen if the resources existed to reduce the probability of involvement in future firesetting incidents.

Johnny's firesetting appears to be the result of an inventive mind. Curiosity led him to light the chemical solution in the test tube, although he did have to search for his mother's matches. He was playing with fire without parental supervision. If Johnny or his parents were able to extinguish the fire themselves, then it is likely that his parents would reprimand and/or punish him because of his behavior. Johnny's parents may or may not review fire safety rules at home and caution him against playing with fire, especially without adult supervision. If the fire department were called to extinguish the fire, one of the firefighters is likely to give Johnny a good, tough lecture about playing with fire.

Ideally, although Johnny's firesetting is clearly the result of accidental play, his age indicates that he should have known about the consequences of playing with fire. Both the parents and Johnny need to be educated about fire safety in the home. The parents need to be taught how to communicate appropriate skills so that Johnny will understand the dangers of fire and not be tempted to overlook them in favor of exploring his environment. In addition, Johnny might benefit from learning fire safety rules for himself and his family. Educational materials and one or two trips to the fire department might help him understand the positive as well as the negative effects of fire. It is estimated that if children playing with fire out of curiosity receive immediate educational intervention, the probability is virtually zero that they will become involved in future firesetting.

Donna's firesetting behavior has gone beyond the curiosity phase, and her repeated episodes are cause for serious concern. To date, none of her firesetting incidents have resulted in any significant damage, including the last oil-burning episode on the stove. In addition, because Donna's mother is physically disabled and extremely dependent on her, she is likely to downplay the serious nature of her daughter's firesetting. Donna's firesetting can be viewed as a cry for help and may represent her frustration at having to grow up so quickly and assume so much adult responsibility at so young an age. In all likelihood, Donna's mother will express her anger over yet another firesetting incident, but because of the dependency between her daughter and herself, she is not likely to take any further action to remediate the behavior. Under these circumstances the probability is high that Donna will be involved in future firesetting episodes that may result in more serious and damaging fires.

Because of the complex relationship between Donna and her mother, there appears to be underlying psychological and interpersonal conflicts that need attention. Donna, out of anger and frustration, has turned to firesetting, to call attention to her own turmoil. Ideally, Donna and her mother could benefit from a psychological evaluation and recommendation for treatment. Of immediate concern is elimination of the firesetting behavior, and longer-range treatment should be focused on making the relationship between Donna and her mother work more smoothly and be more satisfying to both of them. If Donna's cry for help is answered, her firesetting is likely to stop, and if the conflicts between Donna and her mother are addressed, the likelihood of her becoming involved in future, aggressive, acting-out behaviors (such as firesetting) will be greatly reduced.

Gary's pattern of firesetting is the most firmly established of these three youngsters and has already caused significant changes in his life-style. After Gary's most serious firesetting incident, the burning of his grandfather's barn, he was sent to a residential home where a structured program was expected to modify his behavior. The residential home was not the appropriate placement for Gary and, in fact, probably exacerbated his firesetting tendencies. However, the end result, as viewed by juvenile court, is yet another "arson" crime for which Gary can be convicted and should be punished. This time there is no question in the judge's mind that Gary should be incarcerated in a juvenile jail facility to serve "time" for his criminal activities.

Gary demonstrates an unexplained fascination with firesetting which coincidentally emerged with his entry into adolescence. Although his initial firesetting episodes may or may not be viewed as accidental, the fires set in the residential home and school gymnasium appear to be intentional. Juvenile court tried a "lesser" punishment for his first offense, but when Gary was presented as a repeat offender, the court decided that a more severe incarceration experience was appropriate. Unfortunately, after Gary's incarceration, he will be back on the streets again, and neither he nor his family will have understood what motivated the firesetting. In addition, the probability is high that Gary will become involved in future delinquent activities which could include firesetting, but also may involve other antisocial activities. Therefore, it is critical that at this early point in Gary's predelinquent career, he and his family get the type of help they need in order to understand how to redirect Gary toward a more productive life-style. At the very least, Gary needs a comprehensive psychological evaluation and a long-term, intensive treatment plan to both remediate the firesetting behavior and help him through what is predicted to be a somewhat stormy adolescence. Gary, with the support of his family, is likely to respond positively to this type of help and gain more control and direction over his life.

In the "worst" possible scenarios, these three youngsters either will receive no help for their firesetting behavior or, as in the case of Gary, will be incarcerated or punished for their activities. In the "best" possible scenarios, these youngsters and their families will receive appropriate interventions, which may include educational activities, psychological evaluation, psychotherapy, or some type of community-based treatment. This brings us to the point of asking about the current state of the art in the evaluation and treatment of youthful firesetters. What is the clinical profile of youngsters involved in firesetting? What are the current methods for interviewing and evaluating youthful firesetters and their families? What types of intervention strategies are available to eliminate firesetting behavior and the accompanying psychopathology? These questions provide the framework for understanding the current "edge of the field" with respect to the psychology of child firesetting.

YOUTHFUL FIRESETTING—A HISTORY OF THE PROBLEM

Youngsters like Johnny, Donna, and Gary stimulated researchers and clinicians from a variety of fields, such as education, mental health, medicine, social work, criminology, and the fire service, to address the problem of youthful firesetting. As early as the 1940s studies were published describing youngsters involved in firesetting (Yarnell, 1940; Lewis & Yarnell, 1951). Although little knowledge was available as to the incidence or prevalence of the problem, these early researchers identified specific personality characteristics that provided the cornerstone for the current methods of describing youthful firesetters. Although there was a relatively early start in delineating the youthful firesetting syndrome, it was not until statistics on the increasing incidence of the problem emerged in the 1970s that a major multidisciplinary effort was undertaken to examine the etiology and treatment of firesetting. At first, attention was focused on designing a classification system to detect the severity of firesetting behaviors in children and adolescents. Subsequently, recent work has focused on developing intervention strategies aimed at eliminating youthful behavior and remediating the associated psychosocial determinants.

Detection

In 1975 at the California State Psychological Association convention, a panel of psychologists and fire service representatives met to discuss the problem of youthful firesetting. It was concluded that many fire departments as well as psychologists were referred children and adolescents exhibiting firesetting behavior, but most were uncertain how to evaluate and help these youngsters. As a result, a Fire Service and Arson Prevention Committee was formed for the purpose of developing methods to work with children involved in firesetting. A grant from the United States Fire Administration was awarded for the project. The project leaders, Captain Joe Day of The Los Angeles County Fire Department and Kenneth Fineman, Ph.D., focused on the development of a method to classify the severity of youthful firesetting behavior and assess the risk of future involvement. This ground-breaking work provides the foundation for evaluating and classifying youngsters at risk for firesetting (FEMA, 1979, 1983).

This method of describing youthful firesetting behavior was primarily designed to be used by fire department personnel. From interview questions and a brief written questionnaire, using children and their parents as informants, the youngster's level of risk for becoming involved in firesetting is evaluated. There are three major levels of risk—little, definite, and extreme—each representing successively more severe degrees of firesetting. Many factors go into determining firesetting risk, including family history, the child's physical and psychological makeup, and previous and current firesetting incidents. In general, the little-risk level represents children who have set one or two fires out of experimentation or curiosity; the definite-risk level describes children who have been involved in repeated firesetting episodes that may reflect underlying psychological and social conflicts; and the extreme-risk level designates youngsters whose repeated firesetting is a signal of serious behavioral or emotional disturbances. This system of detection and classification is important as an initial evaluation tool for determining the severity of individual cases of firesetting. It is particularly useful to fire department personnel, who often are the first to be referred and subsequently interview youthful firesetters.

Intervention

Children and adolescents benefit from this initial screening and evaluation, but once a determination of risk level has been assessed, then recommendations for intervention become critical. Utilizing the broad categories of little, definite, and extreme risk, a typology of intervention strategies has been suggested corresponding to each risk level (Gaynor, McLaughlin, & Hatcher, 1983). For those children classified as little risk, an educational approach has been shown to be most effective in reducing future firesetting involvement. For definite-risk children, a complete psychosocial evaluation is recommended accompanied by one or more interventions such as counseling, psychotherapy, or help from various community services to meet the individual needs of the youngsters and their families. Extreme-risk cases, although occurring relatively infrequently, not only require a complete psychosocial assessment, but also necessitate long-term, intensive inpatient or residential treatment. This typology of intervention strategies represents general treatment recommendations that are used in conjunction with the system of classifying the risk level of youthful firesetters. These methods of detection and intervention represent the earliest attempts at developing an evaluation-and-treatment package to be used by frontline personnel, such as firefighters and arson investigators, who often are first to come into contact with firesetting youngsters. These methods of detection and intervention have received wide acceptance and application within fire departments across the country.

This pioneering work initially conducted to develop screening and intervention methods for fire departments stimulated mental health care professionals to take a closer look at the problem of youthful firesetting. As a result, recent work has focused on refining methods of evaluating youthful firesetters and their families and designing a variety of intervention strategies to eliminate the problem of firesetting and remediate the accompanying psychopathology. In addition, new roles, functions, and responsibilities have emerged for mental health care professionals with respect to carrying out a common set of goals to reduce the incidence of youthful fireplay, firesetting, and juvenile-related arson.

YOUTHFUL FIRESETTING—AN APPRAISAL

Current work focused on detection and intervention represents significant advances in understanding the psychology of child firesetting. The intention is to present these accomplishments as they represent the two major areas of detection and intervention, describe their current approaches and methods for working on the problem of youthful firesetting, and evaluate the relative merit of these techniques in helping to eliminate nonproductive firestarting and the concomitant psychosocial determinants that disrupt the lives of youngsters and their families. In addition to appraising the current state of the art of detection and intervention methods, new concepts and theoretical models will be proposed to provide an organizational framework for the knowledge that is accumulating in the attempt to understand why youngsters firestart, how they can be helped to stop, and how to prevent this destructive and devastating behavior. Mental health care professionals can assume the leadership in developing and implementing a sound technology of detection and intervention. Consequently, they will take on new roles, functions, and responsibilities as they begin to approach successful solutions to the problem of child firesetting.

Detection

The topic of detection refers to methods of describing and explaining the determinants of youthful fire behavior. At least three current perspectives or systems of thought provide operational definitions of fire behavior. These three perspectives—legal, psychosocial, and clinical—offer different explanations of fire behavior based on the application of the behavior within their respective operational systems. For example, the legal definition of fire behavior specifies the parameters of the crime arson for which youngsters can be arrested, tried, convicted, sentenced, and incarcerated. In contrast, the psychosocial definition of fire behavior details the individual characteristics, social circumstances, and environmental conditions that contribute to the prediction of the occurrence of fireplay and firesetting behaviors. And finally, the clinical definition of fire behavior outlines the relationships between the psychosocial determinants of the behavior and the specific types of psychopathology that accompanies youthful firesetting. Hence, the operational definition of fire behavior and the corresponding delineation of its determinants depends, in part, on the particular system of thought being utilized to describe and explain youthful firesetting.

The legal definition of arson incorporates mental elements with the criminal intent to commit arson. The origins of the legal definition of arson reside with the classification of adult firesetting behavior, because the statutes on arson are written primarily for the adult offender. Therefore, the relationship between adult firesetting behavior and the crime of arson strongly influences the definition of youthful arson crime. A determination of whether youngsters participate in the crime of arson takes into account their motive as well as whether they were mentally responsible for their behavior. In order to classify firestarting as arson, motives must be normal (as opposed to morbid) and the elements of mental responsibility must be evident in the firesetting behavior. The criminal-legal system consists of five phases and associated procedures that are designed to work with youthful firesetters. The concepts of motive and intent that comprise the legal parameters of arson, coupled with the five phases and associated procedures of the criminal-legal system, are the essential features characterizing how the law defines fire behavior and how its intervention methods work with youthful firesetters.

The psychosocial definition of youthful fire behavior involves mapping the relevant determinants, dimensions, and variables that contribute to the prediction of fire behavior. The three major psychosocial determinants of fire behavior are individual characteristics, social circumstances, and environmental conditions. Each of these psychosocial determinants is comprised of dimensions that are operationally defined by quantitive variables. For example, one of the dimensions of individual characteristics is demographic. It is defined by the variables of sex, age, race, and socioeconomic status. Each of these variables can be measured and the dimensions they define can predict how youngsters will express their interest in fire. Theory suggests, although empirical validation remains to be accomplished, that a certain constellation of psychosocial determinants will predict the occurrence of fireplay, firesetting, and juvenile-related arson behaviors.

The clinical definition of fire behavior rests with detailing the relationship between the psychosocial determinants of fireplay and firesetting and the psychopathology that accompanies nonproductive firestarting behavior. There are critical elements which identify the specific features of this psychopathology and which must be organized in such a way so as to yield an effective clinical evaluation of firesetting youngsters and their families. In addition, a system must be set into place for arriving at a clinical diagnosis utilizing the current classes of major childhood psychiatric disturbances. Identification of the critical elements of psychopathology and determination of a relevant clinical diagnosis comprise the major aspects of the clinical definition of fire behavior.

The legal, psychosocial, and clinical parameters represent three diverse methods of operationally defining fire behavior. The legal system interprets certain kinds of fire behavior as criminal, the psychosocial system predicts the occurrence of a range of different types of fire behaviors, and the clinical system identifies the elements of psychopathology that accompany fireplay and firesetting. A comprehensive presentation of detection must include all three perspectives and their contribution to the description and explanation of the determinants of fire behavior.

Intervention

Once evidence of pathological firesetting has emerged, methods of intervention must be considered to eliminate the nonproductive firestarting behavior and the accompanying psychopathology. In addition to exhaustive psychosocial evaluation, the major classes of intervention are psychotherapy and community-based programs. A successful treatment outcome for youthful firesetters and their families depends on an optimal match between the accurate identification of specific behaviors targeted for change and the selection and application of one or more appropriate intervention strategies.

Recently, there has been an emphasis on developing psychotherapies to work specifically with youthful firesetters and their families. A variety of outpatient and inpatient treatment approaches are designed to eliminate nonproductive firesetting and remediate the accompanying psychopathology. Cognitive-emotion, behavior, and family psychotherapy represent the three predominant methods of outpatient therapy. Psychodynamic and behavior therapy are the major methods of inpatient intervention. Although case studies, and in a few instances follow-up investigations, suggest the effectiveness of these psychotherapies, there is an absence of controlled clinical studies verifying the relative merits of these intervention strategies. Nevertheless, preliminary work indicates that these methods are clinically successful in eliminating recurrent firesetting behavior and remediating the associated psychopathology.

In addition to the development of psychotherapy methods to treat youthful firesetters and their families, there has been an effort to implement community-based intervention programs to identify, educate, counsel, and establish referral linkages for youngsters and their families presenting with pathological firesetting behavior. The two types of community programs are prevention and early identification and intervention. Prevention programs, conducted primarily in school settings, operate under the assumption that fire interest is a curiosity that naturally occurs in children. Therefore, an effort must be made to educate all youngsters in fire safety. Early identification and intervention programs, operated primarily by fire departments, offer evaluation, education, counseling, and referral services to firesetting youngsters and their families. As with the psychotherapies, there has been little work accomplished in systematically evaluating the relative effectiveness of these community-based intervention programs, although case reports and some research and development efforts suggest that these strategies show promise in reducing the incidence of fireplay, firesetting, and juvenile-related arson.

Intervention strategies offer the tools with which mental health care professionals can work to eliminate firesetting behavior and remediate the accompanying psychopathology. Subsequent to a thorough psychosocial evaluation, clinicians can fit the specific psychological behavior and environmental conditions targeted for change with the particular set of appropriate intervention methods designed to effect a successful treatment outcome. There is an obvious need for substantial clinical research, but preliminary findings indicate that there are several effective ways to stop nonproductive firestarting and there are a number of psychotherapies and community-based intervention programs that successfully remediate the specific types of accompanying psychopathology. The current menu of intervention strategies offers successful treatment options to youthful firesetters and their families.

Models

Although there has been a rapid growth in the amount of information both in describing and in explaining youthful fire behavior as well as in the development of intervention strategies designed to eliminate recurrent firesetting behavior and the accompanying psychopathology, there have been few attempts to provide an organizational framework for this accumulated knowledge. Two conceptual frameworks are proposed—a predictive and a proactive model—which represent initial efforts to organize current information on the psychology of youthful fire behavior. The predictive model provides operational definitions for the range of fire behaviors, outlines the psychosocial determinants, dimensions, and specific variables hypothesized to predict this range of fire behaviors, and suggests how to proceed with empirically validating these theoretical relationships. The proactive model offers a general intervention strategy designed to reduce the incidence of fireplay, nonproductive firestarting, and juvenile-related arson. The intention is that these models begin to stimulate research activity focused on understanding why youngsters firestart and identify cost-effective methods of eliminating recurrent firesetting behavior and the associated psychopathology.

Roles

Along with the development and implementation of detection and intervention strategies designed to reduce the incidence of fireplay, firesetting, and juvenile-related arson come traditional as well as innovative roles to be assumed by mental health care professionals. The four predominant roles are clinician, consultant, expert witness, and advocate. Each of these roles carries with it various functions and responsibilities specific to working with youthful firesetters and their families. In addition to the successful execution of these professional roles, functions, and responsibilities, mental health care providers can work toward a common set of goals that are likely to reduce the incidence of nonproductive firestarting and make a significant contribution to advancing knowledge in understanding the psychology of child firesetting.

The Current Contribution

The topics of detection, intervention, conceptual models, and professional roles represent the scope of the forthcoming chapters. The intention is to contribute to the understanding of youthful fire behavior by underlining the salient issues and concerns. This effort is the first step in defining the parameters of the field of child firesetting. The current work must stimulate others to take a closer look at the problem and initiate new efforts directed at understanding and working with the complex behavior of child firesetting.

SUMMARY

The problem of youthful firesetting is introduced by describing three case histories of youngsters exhibiting various degrees of mild to serious firesetting behavior. These children represent a hidden majority of youngsters who are both victims and setters of fires. Until recently, youthful firesetting had gone unrecognized as a major, life-threatening social problem. However, because of new and startling statistics revealing the extent and nature of the problem, increasing attention is being paid toward developing methods for reducing the incidence of youthful firesetting. During the late 1970s, pioneering work was accomplished in developing a system for classifying the severity of firesetting behavior and in designing a corresponding typology of recommended intervention strategies to be used in fire departments across the nation. This work stimulated mental health care professionals to take a closer look at the problem of youthful firesetting, refine methods of evaluating firesetting youngsters and their families, and design a variety of intervention strategies to eliminate nonproductive firestarting and the accompanying psychopathology. In addition to describing the menu of current detection and intervention methods and suggesting the need to evaluate their relative effectiveness, conceptual models are presented as a way to organize the accumulating body of knowledge in these areas. Finally, traditional as well as innovative roles are identified for mental health care professionals to assume in their efforts to work toward effective solutions to the problem of child firesetting.

2

Youthful Firesetting and the Crime of Arson

The crime of arson has the highest percentage of juvenile involvement than any other crime indexed by the Federal Bureau of Investigation (FBI, 1985). In 1983, the modal age range for total arson arrest involvement was between 13 and 15 years (Akiyama & Pfeiffer, 1984). In 1984, there were 19,000 arson arrests; 43% of the arrestees were under age 18 and 64% were under 25 years of age, indicating a 7% increase in youthful arson arrests from the previous year (FBI, 1985). This relatively high rate of youngsters involved in the crime of arson necessitates an understanding of how the behavior of firesetting is defined as the criminal activity of arson. Because statutes are developed primarily for defining adult arson behavior, the determination of whether youthful firestarts are arson resides, in part, on the classification system utilized to describe adult firesetting behavior. A typology for categorizing adult firesetting behavior is presented to identify the essential features of adult fire behavior which relate the legal definition of arson to youthful fire behavior. These essential features are motive and intent. If it can be demonstrated that motive and intent sufficiently characterize youthful fire behavior, then youngsters can be arrested, tried, convicted, and sentenced for the crime of arson. However, the predominant question that surrounds youthful fire behavior is whether youngsters under the age of 14 are capable of demonstrating intent or mental responsibility for their actions. The six elements defining intent or mental responsibility are reviewed in terms of how they specifically relate to youthful fire behavior. Finally, when youngsters become identified as firesetters by arson investigators, they are introduced into the criminal-legal system. The phases of investigation, arrest, trial, sentencing, and incarceration that comprise the system of juvenile justice are outlined for the purpose of understanding the potential path that lies ahead for youngsters once they participate in an intentional, nonproductive firestart. The question of effective deterrents to the crime of arson is raised in light of what is currently viewed as the standard intervention method employed by the system of juvenile justice.

THE LEGAL DEFINITION OF ARSON

The legal definition of arson is based on the classification of adult firesetting behavior. This is because statutes are developed primarily for defining adult arson behavior. The determination of whether a youngster's firestart is an arson crime resides within the criminal-legal parameters of adult arson behavior. Therefore, it is important to understand the origins of the legal definition of arson as they relate to adult firesetting behavior. Although there are federal and state laws defining the crime of arson, the state statutes provide the most explicit definition of the criminal-legal behavior of arson. It has been interpreted that the mental state in arson defines the crime of arson (Pollack, 1980). Motive and intent play a critical role in determining whether a youngster's firestart can be classified as arson. Hence, it is necessary to distinguish between the concepts of motive and intent and to define their characteristics as they relate to demonstrating whether a youngster's firesetting is a crime of arson. A set of specific conditions must exist before firestarting can carry the criminal-legal label of arson.

Origins of the Legal Definition

There is a significant literature from a variety of disciplines which describes the psychology of adult firesetting behavior. Perhaps the most recent and most widely accepted classification of firesetting behavior identifies 11 major types of adult behavior (Rider, 1980). This system of classification is utilized as a guide by law enforcement officials and arson investigators to identify and apprehend the adult arsonist. This typology will be described to offer a perspective on the interface between adult firesetting behavior and the criminal-legal definition of arson. In addition, the evidence will be reviewed indicating whether linkages exist between adult arson behavior and child firesetting. Although the criminal-legal definition of arson is based on the classification of adult firesetting behavior, there are preliminary indications that similar behavioral characteristics may be present in adult and child firestarting activities. Regardless of the age of the offender, the statutes define the legal conditions that identify the crime of arson.

There are 11 typologies currently classifying the adult firesetter (Rider, 1980). They are: (1) the arrested arsonist, (2) the incarcerated adult arsonist, (3) the paroled adult arsonist, (4) the revenge firesetter, (5) the jealousy motivated firesetter, (6) the attention-seeking firesetter, (7) the volunteer fireman solitary firesetter, (8) the fire ''buff'' firesetter, (9) the excitement firesetter, (10) the psychotic firesetter, and (11) the pyromaniac. Several of the categories, such as the arrested, incarcerated, and paroled arsonists, are self-explanatory. Many of the categories were developed utilizing the motive or reason for firestarting as the major dimension of classification. For example, the revenge, jealousy motivated, and attention-seeking firesetters represent this type of behavioral category. A few of the typologies are comprised of less frequently occurring, special circumstance classifications such as the volunteer fireman who sets a fire to put himself to work; the fire ''buff'' who spends most of his leisure time at the fire station and often responds to alarms; and the excitement firesetter who sets a fire because of the experienced arousal from the emergency response as opposed to the actual fire.

The remainder of the classifications, the psychotic firesetter and the pyromaniac, reflect abnormal mental states. The psychotic firesetter category is represented primarily by the diagnosis of schizophrenia. The pyromaniac is recognized as a separate diagnostic category as is characterized by a recurrent failure to resist impulses to set fires; an increasing sense of tension before setting fires; an experience of intense pleasure during the fires; and a lack of normal motive such as monetary gain or adherence to a sociopolitical ideology (DSM-III, 1980). These 11 typologies represent a comprehensive classification system for describing adult firesetting behavior.

The 11 categories identifying adult firesetting behavior are not mutually exclusive; rather, they serve to encompass the domain of potential factors characterizing adult fire behavior. Some of these factors associated with adult firesetting can be identified with youthful firestarting behavior. Although there have been few studies examining the psychological and behavioral linkages between child and adult fire behavior, there is preliminary work suggesting that specific firesetting behavior exhibited by adult arsonists may have first been evidenced during their childhood firestarts (Gaynor, Huff, & Karchmer, 1986). For example, adult arsonists reporting childhood firesetting incidents indicated that both during their youthful firestarts as well as during their arson crimes they intentionally watched their fires burn and did not go for help to extinguish them (Gaynor, Huff, & Karchmer, 1986). Therefore, the specific behaviors that accompany firesetting, regardless of the age of the firesetter, may be important factors in determining how to describe and classify firestarting incidents.

The 11 categories characterizing adult firesetting behaviors and preliminary evidence identifying behaviors that may accompany both child and adult firestarting suggest certain features may be critical in classifying firesetting as arson. The typology of adult firesetting behavior indicates that motivation and intention, or the mental state of the firesetter, may be an important factor in describing adult fire behavior. In addition to the psychological or mental state, the specific behaviors accompanying firestarting incidents, such as whether firesetters watch their fires burn and do not go for help to extinguish them, may indicate whether firesetting can be classified as arson. The criminal-legal definition of arson incorporates many of the essential features of adult firesetting behavior.

The Statute

Federal and state laws provide the criminal-legal definition of arson. State statutes outline the specific circumstances that classify firestarting behavior as arson. The implementation of the laws defining the felony crime of arson vary from state to state; therefore, as a point of reference, the state of California's Penal Code will be used as a representative example of how the statutes define arson. California Penal Code, section 451 states:

> A person is guilty of arson when he willfully and maliciously sets fire to or burns or causes to be burned or who aids, counsels or procures the burning of, any structure, forest land or property (California Penal Code, 1979).

The two key terms in this statute are maliciously and willfully. Maliciously refers to a wish to vex, defraud, annoy, or injure another person. Willfully refers to the mental state of the firesetter. It has been suggested that mental state basically defines the crime of arson (Pollack, 1980). This mental state is identified as the criminal or illegal will or intent to set fire or burn. The term intent signifies a purpose or plan that motivates, directs, and moves the individual to firestart. Hence, it is the nature (malicious) of the purpose (intent), in addition to the act of firestarting, which defines the crime of arson.

The criminal-legal terms in this definition of arson must be clarified with respect to how they relate to youthful fire behavior. Because of the critical role motive and intent play in the legal determination of whether a firestart can be classified as arson, it is essential to operationally define a youngster's mental state of motive and intent. The law primarily is concerned with proof of intent or mental responsibility; proof of motive is pursued to strengthen the reasons of intent. Therefore, it is important to distinguish between the concepts of motive and intent and to define their characteristics as they relate to demonstrating whether a youngster's firestart can be classified as a crime of arson. The delineation of the elements comprising motive and intent or mental responsibility is particularly significant because in many states, including California, the Penal Code indicates that youngsters under the age of 14 can be convicted of a crime if it can be proven that they know and understand the wrongfulness of their actions (California Penal Code, 1981).

Motive

The criminal-legal definition of motive consists of the mentalistic, causal driving force or reason that leads a youngster to form the intent for and the execution of the criminal act of arson (Pollack, 1980). A youngster's reason or motive for firesetting will determine whether the act can be classified as the crime of arson. If the motive is "normal," then the probability increases that the youngster's firestart is arson. If the motive is "morbid," that is, the youngster's reasons for firesetting reflect substantial emotional immaturity or are indicative of psychopathology, then the probability decreases that the act of firestarting is arson. Therefore, a determination of whether a youngster's reasons for firesetting are normal or morbid will contribute to the decision of whether an act of firestarting can be classified as an arson crime.

Normal motives for a youngster's firesetting are defined for criminal-legal purposes as a conscious set of deviant reasons which represent the expression of conflict or emotion (Pollack, 1980). Four major reasons are cited as normal motives classifying a youngster's firestart as arson:

(1) Arson is committed if firesetting accompanies another crime, such as vandalism, or covers an accompanying crime such as burglary.
(2) If the firestart is the result of fun or malicious mischief, the act is arson.
(3) A lack of concern about the serious consequences of a fire is evidence of arson.

(4) If the firesetting behavior is motivated by the emotional expression of revenge, anger, or spite that appears understandable or warranted under some circumstances, the resulting fire is the execution of the crime of arson.

Therefore, if a youngster's firestart represents a conscious behavior motivated by one of these four reasons, the chances are high that the criminal-legal system will label the firesetting behavior as arson.

Although a youngster can be accused of arson because firesetting is the result of normal motives such as the emotional expression of revenge, anger, or spite, these affective responses may also represent psychopathology. If these affective responses reflect significant psychopathology, then a youngster's firestarting can be attributed to morbid motives. Therefore, it is necessary to distinguish between the affect of revenge which is indicative of a morbid motive and suggests psychopathology and one which is not. There are specific features that characterize morbid affects. In general, a morbid affect can be aroused more easily; it is expressed inappropriately, exaggerated in intensity, and unduly prolonged; and it is maladaptive, malresponsive, idiosyncratic, and peculiar to the situation. For example, the expression of a morbid affect is illustrated by a case in which a youngster set the back porch on fire in anger because his mother had refused permission for an outing and instead had left the house requesting that the youngster take care of a younger sibling. When questioned about the resulting fire, which burned the house to the ground and from which the youngster initially was perceived as a "hero" for saving the life of the younger sibling, the youngster admitted to feeling intense anger and abandonment immediately prior to the firestart. In addition, the youngster reported the desire to gain recognition and attention through firesetting from his mother as well as from the community. Hence, in this case, the youngster's anger was easily aroused, exaggerated in intensity, and expressed inappropriately through participation in firestarting. The expression of this morbid affect reflects the presence of psychopathology and suggests that the youngster's firesetting may have been the result of an unconscious and irrational motive. Moreover, there appears to be little, if any, understanding on the part of the youngster as to the wrongfulness of the firestarting action.

In addition to the presence of morbid affects, a youngster's firesetting behavior may be attributed to other types of morbid motives. Some of these morbid motives include the desire to obtain a feeling of mastery and power by firestarting and fire extinguishing, the desire to achieve sensual or sexual satisfaction by watching fire burn, and the need to release anxiety and tension resulting from significant, stressful life events such as death of a loved one, major changes in family structure, or recent geographical moves. Although the expression of morbid affect or the presence of morbid motives does not necessarily excuse the youngster's firestart from being classified as arson, they raise important questions regarding whether a youngster's firesetting behavior was motivated by a conscious, rational decision to burn or destroy by fire.

Intent

An examination of motives must be coupled with an analysis of intent to determine whether a youngster's firestart is arson. Whereas motive refers to the reason for firesetting, intent includes both the purpose or design that motivates as well as a description of the mental state leading up to, during, and immediately subsequent to the firestart (Pollack, 1980). There are a number of psychological elements defining a youngster's mental functioning which demonstrate intent to illegally set fire. These mental functions are inferred from a youngster's self-reported introspective observation and description of behavior and from the evidence as observed and reported by others of the characteristics surrounding the act of firesetting. To classify a firestart as arson, not only must there be a normal motive for the act, but it must be demonstrated that a youngster was mentally responsible for the behavior.

A youngster must have intended and participated in the act of firesetting with a mental state that is characterized as being "sound" and "sane" in order to demonstrate mental responsibility and thereby classify the firestart as a criminal act of arson (Pollack, 1980). Six major elements comprise the definition of mental responsibility. The youngster must be capable of all six of these elements in order to have the mental capacity for the intent of arson.

The first of the six elements characterizing mental responsibility requires that the youngster demonstrate consciousness at the time of committing the firestart. Consciousness is defined as the mental state of alertness, accompanied by the capacities of sensation and perception and the ability to appropriately interpret the external stimuli in the environment. In the case example previously presented, wherein the youngster set fire to the back porch of the house because of feelings of intense anger and rejection, although there may have been an overwhelming experience and expression of emotion, there appears to be no evidence indicating a lack of consciousness or loss of perception regarding the surrounding environment. A youngster demonstrating loss of consciousness would report a loss of control over the act of firestarting, which may include an inability to remember the events leading up to the firesetting behavior.

The second element of mental responsibility indicates that a youngster must have the capacity to exercise free will. A youngster must be able to demonstrate the ability to voluntarily and purposefully choose a course of conduct from among a variety of alternative paths of action. In addition, a youngster must be able to consider different options for behavior. In the case of the youngster setting fire to the back porch as a result of experiencing strong feelings of anger, there appears to be no evidence of an attempt on the part of the youngster to consider alternative methods for expressing feelings. Although the youngster may have the capacity to consider options for behavior, during the time immediately preceding the firestart no such consideration was apparent. Therefore, a question remains as to whether the youngster's firestart was indicative of the behavior of free will and thereby represents mental responsibility or criminal intent to burn.

The third element of mental responsibility refers to a youngster's capacity to understand the purpose of the act of intentional firesetting. There are two related components to understanding purpose. First, the youngster must be able to demonstrate knowledge of the physical characteristics of firestarting. That is, the youngster must be able to strike a match or participate in some other manner in igniting the fire. Second, the youngster must understand the physical consequences of firesetting behavior, that is, the potential to harm and destroy objects, property, and persons. In the case of the youngster setting the back porch on fire, it is clear that the physical capability existed for lighting the fire. What is unclear, especially prior to the firestart, is whether the youngster considered the potential physical consequences of the firesetting behavior. Although the youngster may have the capacity for such mental functioning, it may not have been exercised during the experience of emotional intensity preceding the firestart. The evidence suggests that the youngster wanted to gain recognition and attention from setting the fire; however, there appears to be a lack of consideration as to the physical destruction that would accompany the firestart. Hence, both the ability to physically set the fire as well as to understand the consequences of fire together determine whether a youngster is cognizant of the purpose of firesetting.

The fourth element of mental responsibility emphasizes the youngster's ability to participate in goal-directed behavior. Evidence must exist showing that the youngster is capable of developing a plan and carrying out the specific behaviors associated with the plan. The youngster who set fire to the back porch was capable of carrying through with the plan of expressing emotional turmoil. Regardless of the type of planning process, a youngster's ability to plan and subsequently participate in firesetting confirms the ability to engage in goal-directed behavior and demonstrates this element of mental responsibility.

The fifth element of mental responsibility is the youngster's capability for rational thinking. A youngster's rational thinking is demonstrated by participation in a decision-making process that specifies the circumstances of the who, what, when, and why of firestarting. A youngster must exhibit usual and customary reasoning preceding and following firesetting. The critical question surrounding the youngster setting fire to the back porch is whether, because of the overwhelming experience of emotion, mental functioning was impaired to the degree that the accompanying behavior, namely firestarting, was the result of an irrational state of mind. The capacity for rational reasoning and judgment is a necessary requisite for determining whether a youngster is mentally responsible for intentional firesetting.

The sixth, and final, element comprising mental responsibility is whether the youngster understands and appreciates the criminality or illegality of the act of firesetting. It must be demonstrated not only that a youngster is consciously aware that participating in firesetting behavior violates the laws of society, but that the youngster is fully capable of controlling behavior and otherwise conforming to these laws. It is unlikely that before the youngster set fire to the back porch consideration was given to the question of whether firestarting breaks the law, and it is equally unlikely, given the youngster's emotional state, that the firesetting behavior was controllable. An appreciation of the criminality of firestarting is a necessary condition to establish a youngster's mental responsibility and to classify firesetting as an arson crime.

All six elements of mental responsibility—consciousness, voluntariness, understanding of physical characteristics and consequences, goal-directed behavior, rational thinking, and appreciation of criminality—must be evident to demonstrate that a youngster is mentally responsible for an intentional act of firesetting. In the case of the youngster setting fire to the back porch, it is clear that he was conscious and capable of goal-directed behavior; however, because of the overwhelming emotional state, the youngster did not exhibit rational thinking by considering alternative means for expressing feelings, nor was consideration paid to the physical consequences of firestarting or to the fact that firesetting violates the laws of society. Given this somewhat general analysis, it is difficult to confirm the mental responsibility of this youngster as it relates to the firesetting behavior. Therefore, there exists reasonable doubt as to whether this particular act of firestarting was committed by the youngster with criminal intent and can be classified as arson. For this youngster, as for all youngsters participating in firesetting behavior, the six elements of mental responsibility must be present before firestarting can carry the label of arson.

Firesetting Defined as Arson

Table 2.1 shows the two critical components comprising the legal definition of arson—motive and intent—and the corresponding elements that operationally define the associated behaviors. The elements of this criminal-legal definition of arson relate specifically to those behaviors which are likely to characterize youthful motive and

Table 2.1
Firesetting Defined as Arson

Component	Element
I. Motive	A. Accompanies or covers additional crimes
	B. Fun or malicious mischief
	C. No concern of consequences of fire
	D. Expression of affect
II. Intent	A. Conscious
	B. Voluntary
	C. Purposeful
	D. Goal-directed
	E. Rational
	F. Willful

intent for firesetting. If at least one of the four elements of motive is evident and all six elements of mental responsibility can be demonstrated, then the probability is high that the youngster's firestarting can be classified as the crime of arson. However, it is difficult, especially with youngsters under the age of 14, to verify motive as well as mental responsibility for firesetting behavior. In fact, it has been challenged that the majority of youngsters under 14 cannot participate in the crime of arson because their motive is undetermined, their firesetting is not committed with intent (as defined in criminal-legal terms), and consequently they cannot be held mentally responsible for their behavior (Committee on the Judiciary, 1985). Perhaps a more justifiable position is that the questions of motive and intent or mental responsibility must be vigorously pursued for each youngster as they relate to the specific act of firestarting. If a comprehensive analysis reveals some doubt with respect to a youngster's motive or mental responsibility, it is in the best interest of the youngster to delay the decision to label the firesetting behavior as criminal.

THE JUVENILE JUSTICE SYSTEM

There is a systematic process within the criminal-legal system designed to work with firesetting youngsters. In general, this system is a five-phase process beginning with the investigation of the firestarting incident and moving through the mechanisms of arrest, trial, sentencing, and incarceration. There are various procedures describing the operation of each of these five phases of the criminal-legal process. This presentation of the criminal-legal system's treatment of youthful firesetters is based on that which is currently operational in the state of California. These systems and their procedures tend to vary from state to state and differences also are apparent between communities within states. Nevertheless, a description of how the criminal-legal system within the state of California works with youthful firesetters is representative of a fairly acceptable standard of practice.

Investigation

If a youngster is suspected of a firestart that resulted in a fire significant enough to involve firefighting personnel, then the fire department may assign an arson investigator to determine the extent of the youngster's involvement. Arson investigators, firefighters, and police are most likely to encounter youngsters involved in what allegedly is their first firesetting incident, although these youngsters may be suspected of participating in other firestarts occurring within the community. If a determination is made by the arson investigator that a youngster was responsible for the firestart, and there is no previous record of involvement in firesetting behavior, the arson investigator administers the procedures of counsel and release. Counsel and release refers to the arson investigator talking with the youngster about the dangers of firestarting, describing the criminal-legal procedures such as arrest, conviction, and incarceration that could result if firesetting continues, warning the youngster not to engage in further firestarting, and releasing the youngster from further obligation or commitment. If the arson investigator demonstrates that a youngster has repeatedly participated in firestarting, and the current incident under investigation is a costly and damaging fire, and the evidence strongly indicates the youngster's involvement, then the arson investigator is likely to arrest the youngster for the crime of arson.

Arrest

Police and arson investigators have the authority to arrest a youngster suspected of involvement in the crime of arson. When a youngster is taken into custody for the crime of arson, the police may release the youngster, issue the youngster a citation to appear before the probation officer at juvenile court, or take the youngster in custody to the probation officer. The probation officer may release the youngster, keep the youngster in custody given the existence of special circumstances, or release the youngster to family on a form of house arrest with a promise to appear in court at a later date. The probation officer then investigates the case against the youngster, which consists primarily of reviewing the incident reports of the police or arson investigators and assessing the youngster's prior criminal record. The probation officer may decide to carry the case no further, to release the youngster without charges on probation, or to refer the matter to the district attorney for institution of formal juvenile court proceedings. The district attorney may reject the case as factually or legally unsound, or the district attorney may file a petition to commence proceedings in juvenile court.

Trial

When the district attorney files a petition to begin proceedings in juvenile court, and the youngster is in custody, then the youngster has the right to an immediate hearing to decide whether he must remain in custody pending trial. If the youngster is detained in custody at that hearing, then he has a right to a rehearing in three days, at which time a preliminary case against him must be established through the presentation of evidence. If, after this hearing, the youngster remains in custody, he is entitled to a speedy trial. At the trial, the youngster may decide to exclude the public. California law indicates that the trial proceedings will be conducted in an informal, nonadversary atmosphere (California Penal Code, 1971). At the beginning of the trial the youngster enters pleas. The judge may find the evidence insufficient and discharge the youngster. If the judge is convinced beyond a reasonable doubt that the youngster committed arson, the judge sustains the petition filed by the district attorney. This trial is not a criminal proceeding, and a judge's order sustaining a petition is not a conviction for purposes of a criminal record. However, once the judge determines that the youngster participated in an act of arson, the judge has the power to invoke a variety of methods of retribution or punishment.

Sentencing

If the judge confirms the youngster's involvement in arson, then the judge will read a social study of the youngster prepared by the probation department as well as hear any other evidence that may be relevant. The judge has a wide range of alternatives in the equivalent of sentencing. The judge can declare the youngster a ward of the court or release him on probation. If the youngster is declared a ward of the court, he may be removed from the custody of parents and physically confined or placed in foster care. The maximum length that the youngster is in the custody of the court is limited to the maximum length of the prison or jail term an adult could have received for the same offense. If the youngster is released on probation, he is likely to be released in the custody of parents. The youngster may be ordered to pay restitution. Parents of the youngster may be held civilly responsible for the expenses incurred as a result of their youngster's arson, as well as for court costs such as attorney's fees.

Incarceration

If the judge decides that the youngster is a ward of the court, then the judge also must determine the type of confinement. There are a wide range of facilities in which a youngster can be incarcerated. On the one hand, there are day camps, which offer a maximum number of privileges to a youngster, and on the other, there are prisons, which afford no such freedoms. Many of the less restrictive custodial facilities will not accept a youngster with a background of firesetting. During the youngster's period of incarceration, there are few, if any, rehabilitation efforts aimed at preventing the youngster's future involvement in firestarting or at remediating the underlying psychosocial determinants of the behavior. At the termination of the specific period of sentencing, the youngster is released back into the community either to parents or to a foster care family. This represents the end of the youngster's involvement with the system of juvenile justice, unless, while under the age of 18, the youngster participates in another act of arson or some other type of delinquent behavior.

Criminal-Legal Intervention

Table 2.2 summarizes the five phases and associated procedures of how the criminal-legal process works with youthful firesetters. A youngster's progress through the system can be interrupted and terminated at any one of the five phases of investigation, arrest, trial, sentencing, and incarceration. For example, during the trial phase, if evidence indicates that a 12-year-old youngster was not mentally responsible, in that the youngster did not, at the time of the firestart, understand wrongfulness of actions, then the judge

Table 2.2
Phases and Procedures of the Criminal-Legal System for Working with Youthful Firesetters

Phases	Procedures
I. Investigation	A. Counsel and release
	B. Arrest
II. Arrest	A. Release
	B. Probation
	C. Citation
	D. Petition for trial
III. Trial	A. Hearings
	B. Pleas
	C. Sustained petition
IV. Sentencing	A. Ward of the court
	B. Release on probation
V. Incarceration	A. Day camp
	B. Prison

can rule this youngster incapable of committing the crime of arson (Moscrip, 1986). In this case the judge can dismiss the charge of arson and order the youngster and family to obtain necessary psychological treatment. Hence, in general, criminal-legal procedures have been established to offer a youngster adequate and fair protection under the law. However, it is equally important for the youngster and family to understand their respective responsibilities in preventing participation in behaviors such as firestarting which can result in confrontation with the criminal-legal system.

ALTERNATIVE DETERRENTS

Once a youngster participates in a nonproductive firestart that results in a costly and damaging fire, the probability is high that this youngster will experience one or more phases of the criminal-legal system. If this is the youngster's first known firesetting incident, then an arson investigator is likely to have a talk with the youngster and release him to the family. If the youngster has a prior record of involvement in firesetting or other delinquent behaviors, then it is likely that this youngster will be arrested and tried for the crime of arson. If the evidence suggests that the youngster is mentally responsible for committing the crime of arson, then this youngster can be incarcerated in a correctional facility. Hence, the single behavior of firestarting can trigger an unfortunate chain of events for a youngster and interfere with what otherwise might have been a happy and fulfilling childhood and adolescence.

The question to be asked is whether a youngster's interest in fire must result in a confrontation with the criminal-legal system. The answer is that the majority of youngsters who demonstrate an interest in fire need never become a part of the criminal-legal process. There are effective steps which can be taken so that fireplay, nonproductive firestarting, and juvenile-related arson are not outcomes of a youngster's curiosity about fire. In addition, for those youngsters who do participate in firesetting, there are a number of alternative intervention strategies which successfully eliminate firestarting as well as remediate the associated psychosocial determinants. It is possible for youngsters and their families to take advantage of alternative deterrents which are designed to ensure that what begins as a naturally occurring curiosity about fire does not result in the criminal behavior of arson.

An understanding of why youngsters participate in fireplay and firesetting behaviors from a psychosocial and clinical perspective has led to the development of successful alternative deterrents. Perhaps the most significant deterrent for all youngsters is their participation in fire safety and prevention programs throughout preschool and elementary school. However, prevention is not an effective deterrent once youngsters have expressed their interest in fire by becoming involved in fireplay and firesetting. The early identification of these youngsters is the key to the implementation of successful intervention strategies. Among the most effective deterrents are psychosocial evaluation, psychotherapy, and community-based intervention programs. Youngsters and families who choose to participate in these types of interventions reduce the probability of their future involvement in the criminal-legal system.

These alternative deterrents advocate understanding and remediating the psychosocial determinants that stimulate firesetting behavior. This advocacy position is based on the observation that if the causative factors of firestarting can be ameliorated, then there is little likelihood of the reemergence of the firesetting behavior. The intervention strategies of evaluation, psychotherapy, and community-based programs can be utilized successfully to eliminate firesetting behavior and the accompanying psychosocial determinants and to prevent the entry of youngsters into the criminal-legal system as a result of their committing the crime of arson. Furthermore, these alternative deterrents can work with those youngsters already in the criminal-legal system to prevent the recurrence of their firestarting behavior. Despite dramatic differences in the rehabilitative methods employed by the alternative deterrents and the criminal-legal process, both these intervention systems share the common goal of reducing the incidence of juvenile-related arson.

SUMMARY

The criminal-legal system records arson crime as having the highest percentage of youthful involvement. Because the origins of the legal definition of arson reside in the description and classification of adult firesetting behavior, the essential features characterizing adult fire behavior as criminal, namely motive and intent, also identify youthful fire behavior as arson. For a youngster to commit an act of arson, it must be demonstrated that firesetting was the result of a normal (as opposed to morbid) motive and that the youngster was mentally responsible for the firestart. Normal motives classifying firesetting as arson are if the firestart accompanies or covers another crime; if it is the result of fun or malicious mischief; if there is a lack of concern regarding the serious consequences of a fire; and if it was motivated by emotional expression that appears understandable or warranted under some circumstances. A youngster is mentally responsible if all six elements—consciousness, voluntariness, purpose, goal directedness, rationality, and willfulness—are apparent in the firestarting behavior. The criminal-legal system set up to work with youthful firesetters is comprised of five phases and associated procedures. The five phases are investigation, arrest, trial, sentencing, and incarceration. The youngster's progress through this system can be interrupted or terminated at any one of the five phases, depending on specific criminal-legal procedures. There are alternative deterrents that have been successfully implemented to eliminate firestarting and the accompanying psychosocial determinants and can be utilized to prevent the entry of youngsters into the criminal-legal system.

3

Youthful Fire Behavior—A Predictive Model

To advance the understanding of the psychology of youthful firesetting, it is necessary to propose a conceptual model to organize the emerging theoretical, clinical, and research information. This chapter begins with a critical review of various psychological theories and their supporting clinical and research evidence which attempt to describe and explain youthful fire behavior. Extrapolating from this work, those psychosocial determinants are identified which are hypothesized to predict the occurrence of youthful fire behavior. A conceptual model for defining levels of fire behavior is presented, which suggests that it can be viewed as occurring along a continuum of increased involvement, from fire interest, which represents a natural curiosity about fire, to fire-safe and fire-risk behaviors. Finally, a predictive model of firesetting is proposed which suggests that specific psychosocial determinants predict whether children will be fire-safe and demonstrate fire competence or whether they will become involved in fire-risk behaviors such as fireplay and firesetting. In this predictive model the psychosocial determinants are the independent variables or predictors and the various levels of fire behavior are to be predicted or represent the dependent variables. This predictive model allows for an integrated conceptualization of the many hypothesized variables related to fire behavior and builds the foundation for utilizing specific types of information to predict the occurrence of youthful fire behavior.

HISTORICAL OVERVIEW

There appear to be three primary theoretical frameworks around which the literature on youthful fire behavior is organized—psychoanalytic theory, social learning theory, and dynamic-behavioral theory. Each theoretical perspective postulates why pathological firesetting emerges as a childhood behavior. There are some clinical observations and only a handful of empirically based investigations supporting these theories. What follows is a critical review of the three primary theoretical explanations of childhood pathological fire behavior along with the existing research generated by each of these theories. Special attention will be focused on how these theories and their supporting evidence identify the relevant psychosocial determinants of youthful fire behavior.

Psychoanalytic Theory

The first reports of the behavior of arsonists surfaced in the psychiatric literature in the early 1800s describing a female patient who, as a result of sexual arousal, set fire to her bed (Lewis & Yarnell, 1951). Freud confirmed the relationship between sexual desires and fire by proposing that fire was symbolically expressive of libidinal and strong phallic-urethral drives. He wrote that men attempted to extinguish fires with their own urine, thereby symbolically engaging in a homosexual struggle with another phallus. Freud suggested that the association of sexual feelings with urination was the primary underlying motive associated with the thrill of igniting and then extinguishing fires (Freud, 1932). In support of this analysis, additional psychoanalytic theorists have hypothesized that firesetting is associated with a regression to the urethral-phallic phase of psychosexual development and that the actual setting of fires serves to substitute for forbidden masturbatory desires or to actually excite the firesetter to sexual arousal (Kaufman, Heins, & Reiser, 1961).

There are some observational data to support the psychoanalytic theory of firesetting. Masturbation and orgasm have been reported during firesetting in a small percentage of cases (Lewis & Yarnell, 1951). In addition, sexual dysfunction has been noted in studies of both adolescent and adult firesetting (Stekel, 1924; Heath, Gayton, & Hardesty, 1976). A recent study comparing firesetting and non-firesetting youngsters living in a residential treatment center showed that the firesetting youngsters demonstrated a significantly greater amount of sexual conflict and arousal as measured by their performance on projective tests (Sakheim, Vigdor, Gordon, & Helprin, 1985). Also, enuresis has been linked to firesetting (Lewis & Yarnell, 1951; Kaufman, Heins, & Reiser, 1961; Siegelman & Folkman, 1971). However, there is also some evidence which questions the assumption that sexual conflict is the unique underlying motive of firesetting. First, it has been pointed out that the relatively high incidence of sexual conflicts existing most certainly outweighs the incidence of firesetting (Gold, 1962; Vreeland & Waller, 1979). Therefore, sexual conflict may be related to many pathological behaviors of which firesetting is one. The same argument can be made for enuresis, in that there is a much higher rate of enuresis occurring in the general population as compared to the population of firesetters (Vreeland & Waller, 1979). Hence, it can be concluded that although sexual conflict and enuresis seem to be related to firesetting, they are two factors in a more complex personality style associated with antisocial behaviors.

Social Learning Theory

In the 1960s the popular view of sexual conflict being the primary motivating factor for firesetting gave way to theories which suggested that firesetting may be the direct expression of an aggressive act (McKerracher & Dacre, 1966). It is proposed that firesetting represents a fear of the direct expression of anger (Vreeland & Waller, 1979). The major theoretical underpinnings of this view come from social learning theory, which hypothesizes that the firesetter has experienced a number of social and interpersonal failures and has been generally ineffective in obtaining rewards from his environment. Firesetting becomes a way for the individual to gain some mastery and control over a hostile and unrewarding environment, the kind of mastery and control that heretofore was unobtainable through socially acceptable methods (Vreeland & Waller, 1979).

There are some data supporting the assumption that at least adult firesetting is the result of a misplaced aggressive expression of underlying failures at social and interpersonal life experiences. It should be noted that the most frequently occurring reason for adult arson is revenge (FBI, 1982). In addition, there are a number of studies which suggest that the adult arsonist demonstrates a passive-aggressive personality style and typically displays low self-esteem, possesses few coping strategies, and has a narrow range of interpersonal relationships (Schiller & Jacobson, 1984). Also, there are data indicating that arsonists are more likely to commit crimes against property as opposed to person, suggesting that although conflicts may be of an interpersonal nature, the tendency is to release the aggression indirectly toward inanimate objects (which obviously cannot retaliate) (Wolford, 1972). Although these data support the social learning theory explanation of adult firesetting, they seem to attribute one general dynamic motivation for all adult firesetting. The social learning theory process is intriguing, but it may not be the only plausible explanation for firesetting. To date, there is no research testing the applicability of this theoretical framework as opposed to some other, such as the psychoanalytic perspective. Also, it is difficult to know how the social learning theory explanation of adult firesetting translates to understanding youthful firesetting.

Some social learning theorists have suggested that their model can be applied to youthful deviant behavior (Patterson, 1976, 1978; Patterson et al., 1975). If firesetting can be viewed as one of several specific antisocial behaviors, then social learning theory also can explain youthful fire behavior. The theoretical model contends that aggressive children come from families where one or more members also demonstrate aggressive behaviors. Through modeling, children learn to exhibit aggressive behaviors. As a result, poor social skills begin to develop within the family and continue to occur outside the family, for example, with peers and in school. Hence, the family as well as the youngster's other primary environments reinforce the development of the socially deviant behavior of firesetting.

To date there have been few reported studies confirming the application of social learning theory to youthful firesetting. In support of the social learning theory of deviant behavior, it has been demonstrated that youngsters involved in firesetting also exhibit other antisocial activities such as lying, stealing, and acts of vandalism (Patterson, 1976). Unfortunately, the number of youthful firesetters observed in this study was 12, thereby severely limiting the generalizability of the findings. In addition, there has been some preliminary work suggesting that the fathers of adolescent firesetters all had significant fire-related employment (fireman, furnace stoker, automobile burner at a junkyard), with one youngster not only watching but participating with his father in the burning of automobiles (Macht & Mack, 1968). Hence, there is some preliminary evidence suggesting that modeling may influence the onset of youthful fire behavior. Further work needs to be initiated investigating the family characteristics of youthful as well as adult firesetters to identify the variables associated with the development of aggressive behaviors in general, and specifically firesetting.

Dynamic-Behavioral Theory

Although social learning theory emphasizes the importance of the family and the environment in shaping the development of antisocial behavior, it pays less attention to the predisposing characteristics of the individual. Fineman (1980) has proposed an alternative model that attributes pathological juvenile firesetting to the interaction of a set of personality variables that predispose a child toward firesetting, a specific set of environmental contingencies that teach a child to play with fire, and immediate environmental conditions that motivate the firesetting act. This dynamic-behavioral formulation suggests that the personality variables include historical and constitutional factors. Historical factors consist of family background, specific behaviors exhibited by the child, and school adjustment. Constitutional factors include organic dysfunctions and physical problems. Those environmental contingencies which encourage a child to play with fire are modeling, imitation, and inconsistent negative reinforcement. The immediate environmental conditions motivating firesetting behavior are divided into antecedent and reinforcing events. Antecedent events include stress, emotional distress, and peer pressure. Reinforcing events consist of attention and sensory input. Fineman suggests that personality variables, environmental contingencies, and immediate environmental conditions, and their associated variables, contribute to two major types of juvenile firesetters, the curiosity firesetter and the pathological firesetter. The curiosity firesetter is a young normal child setting fire primarily for reasons of curiosity and environmental exploration. The pathological or recurrent firesetter is motivated by strong emotional distress such as anger or revenge, overriding stress such as death in the family or divorce, and the reinforcing negative attention firesetting will bring from parents, peers, and the community.

Fineman cites numerous clinical studies in support of his theoretical observations. There also appears to be substantial evidence indicating that there are at least two major types of firesetting behaviors—those occurring as the result of children exploring their environment and those occurring as the result of underlying psychological and social problems (FEMA, 1979, 1983). Unfortunately, as Fineman points out, the majority of the studies supporting his theoretical position are clinical in nature and suffer from such basic inadequacies as sampling errors, faulty study design, interviewer, coding, and recording bias, and respondent unreliability. In addition, Fineman does not explain how the three major factors—personality variables, environmental contingencies, and immediate environmental conditions—interact in such a way to specifically elicit firesetting behavior as opposed to other antisocial or delinquent behavior. It would be useful for this theory to suggest how and why these multiple factors combine in such a way to produce pathological firesetting behavior.

From a similar theoretical perspective, Kolko and co-workers (1985) have proposed a conceptualization that identifies several factors within three major content domains which comprise a tentative risk model for firesetting. The three major domains include learning experiences and cues, personal repertoire, and parent and family influences and stressors. Learning experience and cues consist of such variables as early modeling, direct experience with fire, and the availability of firestarting materials. Personal repertoire factors include cognitive components such as limited fire education and safety skills, behavioral components such as interpersonal ineffectiveness, and covert antisocial activities and motivational components. Parental and family influences and stressors are limited adult supervision, parent distance and uninvolvement, parental pathology, and stressful external events. Kolko also recognizes the absence of controlled studies with objective measures to verify this conceptualization. However, he hopes that the model will identify those factors which will diagnose and classify youthful firesetters. He suggests empirical validation will contribute to predicting the emergence of child firesetting behavior.

There seems to be little difference between the theoretical approaches of Fineman and Kolko except for the actual categories and labels they attribute to the variables hypothesized to be related to youthful firesetting. Both theories present a broad-based and inclusive approach to identifying the determinants of youthful fire behavior, and perhaps at this juncture of current research, more specific operational definitions of these factors and inferences on their directionality, are premature. In addition, both theories neglect to detail the interrelationship between psychosocial determinants, although Kolko suggests that not enough data exist to offer ideas regarding these relationships. The dynamic-behavioral theory, as represented by the work of Fineman and Kolko, provides a comprehensive, but as yet unintegrated approach to identifying the psychosocial determinants of youthful fire behavior.

PSYCHOSOCIAL DETERMINANTS

Psychoanalytic, social learning, and dynamic-behavioral theory and their supporting evidence provide clues as to the determinants or predictors of youthful fire behavior. Each of these three theoretical perspectives offers different types of psychological explanations for the emergence of fire behavior, and none of the theories, to date, have generated substantial empirical evidence to reject their specific postulates or hypotheses. Hence, it is important to extrapolate from each of the theories the specific psychosocial determinants hypothesized to predict youthful fire behavior.

Determinants, Dimensions, and Variables

Table 3.1 categorizes three major classes of psychosocial determinants—individual characteristics, social circumstances, and environmental conditions. Each of these three classes of psychosocial determinants is comprised of specific dimensions. For example, the psychosocial determinant of individual characteristics is comprised of the dimensions of demographic, physical, cognitive, emotion, motivation, and psychiatric. These psychosocial determinants and their associated dimensions predict the occurrence of pathological fire behavior in youngsters. Table 3.1 specifies which theoretical framework hypothesizes the relationship of the specific dimensions of psychosocial determinants to pathological fire behavior and cites existing evidence, either clinical observation or empirically based research, to support the theoretical position.

Conceptually, each of the three classes of psychosocial determinants and their corresponding dimensions represent independent variables or predictors of pathological fire behaviors in youngsters. Each of the dimensions can be specified further by variable clusters. Variable clusters operationally define the psychosocial dimensions. For example, the demographic dimension of the individual characteristics determinant consists of the variable clusters of sex, age, and socioeconomic status. Table 3.2 presents the three classes of psychosocial determinants, their corresponding dimensions, and variable clusters which operationally define each of the dimensions.

Individual Characteristics

The psychosocial determinant of individual characteristics is defined as those factors intrinsic to the individual that influence behavior. These characteristics have been referred to as the set of

Table 3.1
Identifying the Psychosocial Determinants of Pathological Fire Behavior

Psychosocial Determinants	Dimensions	Theory	Evidence/References
I. Individual characteristics	A. Demographic	D/B	Lewis and Yarnell, 1951; Fineman, 1980; Gruber et al., 1981; Strachan, 1981; Kuhnley et al., 1982; Stewart and Culver, 1982; Heath et al., 1983; Wooden and Berkey, 1984
	B. Physical	P; D/B	Lewis and Yarnell, 1951; Stekel, 1924; Kaufman et al., 1961; Hurley and Monahan, 1969; Nielsen, 1970; Siegelman and Folkman, 1971; Heath et al., 1976; Sakheim et al., 1985
	C. Cognitive	D/B	Kuhnley et al., 1982; Ritvo et al., 1982; Stewart and Culver, 1982; Kolko et al., 1985
	D. Emotion	D/B	Awad and Harrison, 1976; Block, Block, and Folkman, 1976
	E. Motivation	D/B	Lewis and Yarnell, 1951; Bender, 1959; Kaufman et al., 1961; Award and Harrison, 1976; Gruber et al., 1981; Kolko et al., 1985
	F. Psychiatric	P; D/B	Stewart and Culver, 1982; Kuhnley et al., 1982; Kolko et al., 1985
II. Social circumstances	A. Family	S/L; D/B	Macht and Mack, 1968; Vandersall and Weiner, 1970; Siegelman and Folkman, 1971; Patterson, 1978; Fine and Louie, 1979; Kafry et al., 1981; Gruber et al., 1981; Strachan, 1981; Stewart and Culver, 1982; Ritvo et al., 1982
	B. Peers	S/L; D/B	Vandersall and Weiner, 1970; Patterson, 1978; Heath et al., 1983
	C. School	D/B	Vandersall and Weiner, 1970; Gruber et al., 1981; Kuhnley et al., 1982; Stewart and Culver, 1982

(*Continued*)

Table 3.1
(*Continued*)

Psychosocial Determinants	Dimensions	Theory	Evidence/References
III. Environmental conditions	A. Antecedent stressors	D/B	Fineman, 1980
	B. Behavioral expression	D/B	Patterson, 1978; Fineman, 1980; Kafry et al., 1981; Gruber et al., 1981
	C. Consequences	S/L; D/B	Patterson, 1978; Fineman, 1980; Kolko et al., 1985

P = Psychoanalytic; S/L = Social Learning; D/B = Dynamic-Behavioral

factors that predispose youngsters to pathological fire behavior (Fineman, 1980). These characteristics also can be defined as those attributes which differentiate individuals, or individual differences, which will predict the occurrence of youthful firesetting.

It is proposed that the determinant of individual characteristics is comprised of at least six dimensions. The first dimension is demographic and consists of the variables sex, age, race, and socioeconomic status. The physical dimension includes the variables of general health, abnormal or unusual sexual activities, including abuse, organicity, which may involve such conditions as minimal brain dysfunction, and energy level, which refers to levels of activity, including the diagnosis of hyperactivity. The cognitive dimension consists of a measure of intelligence as well as an assessment of learning ability or whether there exist noticeable learning problems or specific disabilities. The emotion dimension refers to general emotional adjustment and includes the average feeling state as well as the ability to express specific emotions. Motivation is defined as the intention behind the behavior or the stated reason for the firesetting. Finally, the psychiatric dimension is represented primarily by whether there is an accompanying psychiatric diagnosis which suggests the presence of additional behavior disorders. These six dimensions and their associated variables are expected to describe the individual characteristics of youngsters involved in pathological fire behavior.

Social Circumstances

The second psychosocial determinant, social circumstances, is defined as those factors which characterize interpersonal relationships and style of interaction. It is assumed that these factors are skills that are learned initially in the family of origin and subsequently affect future relationships and circumstances. Specific adjustment or performance in social circumstances reflects the ability to establish and maintain significant interpersonal relationships. It

Table 3.2
Defining the Psychosocial Determinants of Pathological Fire Behavior

Psychosocial Determinants	Dimensions	Variable Clusters
I. Individual characteristics	A. Demographic	1. Sex
		2. Age
		3. Race
		4. Socioeconomic status
	B. Physical	1. Health
		2. Sexual activity
		3. Organicity
		4. Energy level
	C. Cognitive	1. Intelligence
		2. Learning ability
	D. Emotion	1. State
		2. Expression
	E. Motivation	1. Intention
	F. Psychiatric	1. History
II. Social circumstances	A. Family	1. Structure
		2. Behavior
		3. Events
		4. Pathology
	B. Peers	1. Interaction
		2. Behavior
	C. School	1. Achievement
		2. Adjustment
III. Environmental conditions	A. Antecedent stressors	1. Events
	B. Behavioral expression	1. Aggression
		2. Attention
	C. Consequences	1. Discipline
		2. Financial
		3. Medical
		4. Legal
		5. Intervention

is hypothesized that these skills may have a negative relationship to the emergence of pathological fire behavior in that social competence and interpersonal interaction may be retarded for youngsters involved in firesetting.

There are three major dimensions comprising the determinant of social circumstances. The first is the family of origin, defined as the family in which youngsters have lived during their early and formative years (ages 1 through 10). The variables describing the family are structure, referring to the type and number of family members, behavior, which typifies the modalities of interaction, and significant family events, such as geographical move, addition of a new family member, death, and divorce. The second dimension of social circumstances is peer relationships. The two primary variables are style of interaction, which refers to methods of communication, and types of behaviors, which refers to the set of mutually shared activities. The third dimension is school experience and consists of the variables of achievement or ability to maintain an average academic standing and adjustment or the ability to behave in socially accepted ways within the school setting. These three dimensions—family, peers, and school—represent the major social circumstances to which youngsters must adjust, and those youngsters who cannot are at risk for becoming involved in antisocial behaviors such as firesetting.

Environmental Conditions

The third class of psychosocial determinants is environmental conditions. These factors are the specific events which occur in the immediate environment, external to individual behavior, but which influence and direct its activity. It is hypothesized that there is an interaction between individual characteristics, the general set of social circumstances, and events occurring in the immediate environment that elicit pathological firesetting behavior. The events occurring in the immediate environment act as a trigger to the firesetting incident. They influence the actual firesetting activity or they influence what occurs subsequent to the firesetting behavior. The probability of environmental conditions alone effecting the emergence of pathological firesetting is low; rather, the stage first must be set by the existence of the relevant individual characteristics and social circumstances before the impact of environmental conditions can be observed.

The three major dimensions of environmental conditions are antecedent stressors, behavioral expression, and consequences. Antecedent stressors can be identified as major life events or changes occurring prior to a firesetting incident. These events do not necessarily have negative connotations, but they often do represent some sort of crisis such as expulsion from school or physical abuse by a family member. Behavioral expression is defined as the need, through firesetting, to express some unrealized feeling, thought, or action. The two variables hypothesized to be related to behavioral expression are aggression and attention. These two variables also are closely related to the individual characteristics dimension of motivation or reason for the firesetting. Finally, the third dimension is the set of consequences resulting from the firesetting behavior. These consequences play a significant role as to whether the pathological fire behavior will recur. The major variables associated with consequences include the disciplinary action taken by parents and supervising adults; the amount of financial loss due to the fire and whether some form of retribution was required; medical involvement or whether there were physical injuries or loss of life sustained in the fire, to either animal or human life; the legal ramifications of the firesetting referring to whether there was an arrest, conviction, sentencing, or incarceration accompanying the incident; and professional intervention, referring to whether educational or psychological programs were engaged in to help remediate the firesetting problem. All three of the environmental factors—antecedent stressors, behavioral expression, and consequences—play a critical role in determining when firesetting will occur and whether an initial incident will evolve into a recurrent, pathological firesetting pattern.

Empirical Validation

The major goal of identifying and defining the three major classes of psychosocial determinants, their related dimensions and variable clusters, is to provide an integrated framework to empirically test the independent variables or predictors of pathological firesetting behavior. The schema presented in Table 3.2 represents a method to empirically evaluate the relationship between the three classes of psychosocial determinants and the occurrence of youthful firesetting behavior. If the relationship of specific independent variables to firesetting can be verified or rejected, then a classification system can be defined that identifies empirically based predictors of pathological firesetting.

Because each class of psychosocial determinants has been categorized into measurable units or variable clusters, the schema outlined in Table 3.2 can be empirically evaluated in a number of different ways. Each variable within a cluster can be investigated, or a number of variable clusters representing different dimensions can be studied. For example, some research has provided data on the sex, age, and socioeconomic status of youthful firesetters. Studies have shown that boys are more frequently involved in firesetting than girls (Lewis & Yarnell, 1951; Fineman, 1980; FBI, 1983) and that their ages typically range from 5 to 16 (Lewis & Yarnell, 1951; Stewart & Culver, 1982). There is conflicting research on the socioeconomic status of those involved in firesetting, with some authors claiming that arson is a white, middle-class crime (Wooden & Berkey, 1984) and other evidence showing a strong representation of the lower income groups (Strachan, 1981; Kuhnley, Hendren & Quinlan, 1982). Hence, within the demographic dimension of individual characteristics, some of the relationships have been specified between the variable clusters of age, sex, and socioeconomic status and pathological firesetting. In this way various dimensions and their related classes of psychosocial determinants can be empirically evaluated. The schema presented in Table 3.2 which hypothesizes the relationship between specific independent variables and pathological firesetting can be rejected, in part, or adjusted and modified according to empirical evidence. Moreover, theories postulating relationships between specific dimensions and firesetting behavior can be investigated (see Table 3.1). Those theories withstanding empirical scrutiny will contribute to the prediction of pathological fire behavior.

Beyond Independent Variables

Although psychoanalytic, social learning, and dynamic-behavioral theory offer substantial information on the determinants or predictors of youthful fire behavior, they pay little attention to conceptualizing and defining the dependent variable, or that which is to be predicted, namely pathological fire behavior. These theories ignore significant evidence indicating that fire interest may be a natural part of the normal developmental process of childhood. If this assumption is true, then the independent variables or predictors of pathological firesetting may be viewed as interrupting or negatively influencing potentially constructive and age-appropriate behavior. In addition, these theories do not adequately differentiate youthful fire behavior from pathological firesetting, nor do they offer specific behavioral descriptions or identify levels of fire behavior. The following section will present a conceptualization of youthful fire behavior which suggests that it can be viewed along a continuum of increased involvement, from fire interest, which represents a natural curiosity about fire, to fire safe and fire-risk behaviors. Different levels of youthful fire behavior will be identified and defined for the purpose of classifying firesetting behavior.

DEFINING YOUTHFUL FIRE BEHAVIOR

A review of previous theoretical attempts at explaining fire behavior shows that the majority of approaches focus on the emergence of pathological firesetting behavior. It might be helpful to back up one step and take a look at the role fire behavior plays in the normal developmental process experienced by the majority of youngsters. Once "normal" fire behavior is defined, then the emergence of pathological firesetting activities can be understood within a framework similar to that of describing other deviant types of behavior.

Age-Appropriate Fire Behavior

It is proposed that youthful fire behavior occurs along a continuum representing increasing levels of involvement with fire depending primarily on the developmental age of the youngster (Figure 3.1). The initial level of experience for children is fire interest, which begins as early as age three. Depending on the influence of psychosocial determinants, fire interest can lead to fire-safe or fire-risk behaviors. Fire-safe behaviors result in youngsters learning to handle fire competently in a supervised setting, whereas fire-risk behaviors can end in serious or damaging fires resulting from either fireplay or recurrent firesetting. Fire-risk and fire-safe behaviors begin to take shape around age six. This conceptualization assumes that fire behavior is part of the developmental skill repertoire youngsters must acquire, equivalent in nature to learning how to cross the street safely, and that specific psychosocial factors can influence the direction and course of the fire behavior.

Fire Interest

There is substantial evidence indicating that most children express a natural curiosity about fire (Block & Block, 1975; Kafry, Block, & Block, 1981). Therefore, fire interest can be viewed as part of the developmental process experienced by the majority of normal

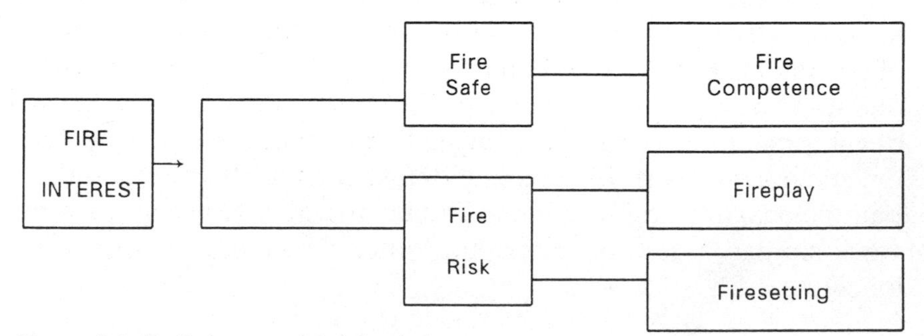

Figure 3.1. Defining youthful fire behavior.

youngsters. Interest in fire can begin as early as age three and may continue until age six or seven. It is most typically observed in young boys, although some girls also are likely to express their curiosity.

Children demonstrate their fire interest in three ways. One of the most typical methods is by asking questions such as "How hot is fire?" and "What makes fire burn?" Usually these questions occur when children observe or are involved in fire-related activities such as watching a campfire or blowing out candles on a birthday cake. A second way in which children express their interest is through play. Youngsters may put on a fire hat, play with their toy fire truck, and pretend they are firefighters racing to put out a fire. Children also play in their toy kitchens and turn their ovens and stove-tops on and off to cook their food. A final way in which youngsters display their interest in fire is by requesting permission to be involved in fire behaviors such as helping to light the barbecue or start the fireplace fire. Fire interest is defined as a natural curiosity about fire and can be expressed through verbal questions, play, and requesting permission to be involved in fire-related activities.

Fire-Safe Behavior—Fire Competence

What begins as a natural curiosity about fire can lead to either fire-safe or fire-risk behaviors. Fire-safe behaviors are defined as those types of activities in which youngsters demonstrate actual competence in using fire materials. Fire-safe behaviors can occur as early as age four. If children express fire interest by asking questions or requesting permission to be involved in fire-related activities, then parents or responsible adults must recognize the interest. If the interest is acknowledged, then youngsters can become involved in activities such as helping to light fireplace fires, birthday candles, or campfires. All of these behaviors can take place under the supervision of parents or a responsible adult. Actual, supervised experience in fire-related activities teaches youngsters to engage in fire-safe behaviors. If youngsters can execute fire-safe behaviors, they demonstrate fire competence. Fire competence represents control and mastery over fire interest and firestarting.

Fire-Risk Behavior

Because fire interest is influenced by several psychosocial determinants, natural curiosity also can lead to fire-risk behaviors. Fire-risk behaviors are defined as those types of activities in which youngsters are accidentally or intentionally involved in an unsafe firestart. Fireplay and firesetting represent the major kinds of fire-risk behaviors. Data show that of those children expressing an interest in fire, 50% actually participate in fireplay (Kafry, Block, & Block, 1981). Fireplay occurs when youngsters experiment with matches or other firestarting materials in an unsupervised setting. Fireplay may or may not result in actual fires. Most fires resulting from fireplay are accidental and unintentional. Following is a case example of a fireplay incident.

> Jason is a seven-year-old who lives in a rural farming area with his 13-year-old brother, Tim, and his mother and father. Jason has successfully finished first grade and will be entering second grade in the fall. Jason's father is a farmer, and Jason and his brother like to help their dad with the farming during the summer months. It is typical for the family to bring their trash to a protected, fenced location. Every two weeks or so, Jason's father or brother burns the refuse. Jason has watched his dad and brother burn the trash, but he never participated in the activity. One afternoon Jason noticed that the trash pile was bigger than usual, so he went to tell his dad or brother. He couldn't find either of them, so, trying to be helpful, he decided to burn the trash himself. He went to the kitchen, just as he had seen his dad do so many times before, and retrieved the box of industrial matches. Jason went out to the trash pile, struck four or five matches, and threw them into the refuse. The trash began to burn in three or four small areas, but very soon the small fires turned into one very large fire. Jason became scared and ran to get help.

Jason is apparently a well-adjusted young boy who likes to help his family run their farm. Unfortunately, Jason was not educated in fire safety, nor was he told specifically that burning the trash was a job that should be left for his older brother or dad. Jason's intention was not malicious, rather he was trying to be helpful. The fire he started was unsafe and he should not have been involved in the activity. Jason's parents should review fire safety rules with him and emphasize that he should not start a fire unless there is an adult present. The consequences of starting a fire which could become uncontrolled also must be reviewed. If these steps are taken it is unlikely that Jason will be involved in another unsafe firestart.

Fireplay can be distinguished from firesetting in that firesetting is the result of an intentional act of a nonproductive firestart. Youngsters are involved in firesetting when they actively seek out fire-starting materials such as matches or lighters and, without adult supervision, ignite papers, leaves, trash, or other flammable objects for such varied purposes as malicious mischief, watching fire burn, revenge, profit, harming person, property, or thing, or other destructive motivations. The resulting fires may be small and easily extinguished or large and require assistance from the fire department. Following is a case example of youthful firesetting.

> Andrew is an eight-year-old child living with his mother, stepfather, and younger, five-year-old sister. Andrew also has a 17-year-old brother, Frankie, who was recently incarcerated in juvenile hall for physically abusing younger children in the neighborhood. Upon questioning by authorities, Frankie also admitted to beating Andrew several times. Neighbors confirmed stories of seeing Frankie throw Andrew to the ground and kick him. Frankie has recently been allowed to make telephone calls home, and plans are being made for his release and return home. Andrew's firesetting began three years ago when his mother found spent matches hidden in his pant pockets. Recently his firesetting behavior increased, and Andrew's stepfather has found several matchbooks hidden in Andrew's room. Andrew's most recent firesetting episode involved him lighting a candle, taking the candle into Frankie's closet, and burning the sleeves off of several of his brother's shirts. Andrew watched the sleeves burn, but told no one

about what he had done. No fire resulted from his activity. Andrew's mother discovered the burned clothes the next day.

Andrew is living in a family where he has experienced neglect as well as physical abuse. His firesetting appears to be recurrent and most recently intentionally aimed at his brother Frankie. Abused children are often unable to react directly and express the anger and aggression they feel toward their attacker. Instead, they typically find other, perhaps more indirect, methods of expressing their feelings of resentment and anger. In Andrew's case his firesetting can be interpreted as an expression of anger, revenge, or retaliation directed toward his brother Frankie. Andrew is unable to express his feelings to Frankie directly, but by burning the sleeves of his brother's shirts he is able to express his hostility in an aggressive and destructive manner. Andrew's intentional firesetting, like that of so many children, is intimately connected with complex psychological problems desperately in need of attention.

Regardless of the psychosocial determinants responsible for youthful fireplay and firesetting, these fire behaviors themselves can be defined by specific behavioral factors. It is necessary to define specific fire behaviors so that their definitions can be utilized as dependent variables in predicting the occurrence of youthful firesetting. Table 3.3 identifies five classes of behavioral factors distinguishing fireplay from firesetting. They are: (1) history of firestart, (2) method of firestart, (3) ignition source, (4) target of firestart, and (5) behaviors occurring immediately following firestart. There are

Table 3.3
Behavioral Factors Distinguishing Fireplay from Firesetting

Factor	Fireplay	Firesetting
I. History	Single episode	Recurrent
II. Method	Unplanned	Planned
III. Ignition	Available material	Flammable or combustible materials
IV. Target	Paper/trash/leaves	Other property
	Own property	Animal/person
V. Behavior immediately following ignition	Put fire out	Watch fire burn
	Go for help	Run away

specific variable clusters associated with each class of behaviors distinguishing fireplay from firesetting. For example, method of firestart is often spur of the moment and accidental in fireplay and planned and intentional in firesetting. If youthful fire behavior can be assessed using these behavioral factors, then fireplay and firesetting can be distinguished and a dependent variable can be developed to empirically test the relationship between psychosocial determinants and youthful fire behavior.

A PREDICTIVE MODEL

A predictive model of youthful fire behavior is being proposed to present an integrated method of conceptualizing as well as quantifying the prediction of pathological firesetting. The underlying assumption of this model is that youthful fire behavior is part of the normal developmental process of childhood. There are age-appropriate youthful fire behaviors beginning with fire interest, which occurs for most children around the age of three. Fire interest can lead to either fire-safe or fire-risk behaviors depending on the influence of psychosocial determinants. Fire-safe youngsters learn to use fire materials in a controlled, supervised setting and can demonstrate fire competence. Fire-risk behaviors are defined as fireplay and firesetting, or represent the category of pathological fire behaviors. Specific psychosocial determinants will predict the occurrence of pathological fire behaviors. Figure 3.2 summarizes this predictive model of youthful fire behavior.

Independent and Dependent Variables

The postulate of this predictive model is that psychosocial determinants, or the independent variables, predict the occurrence of pathological fire behaviors, or the dependent variables. Figure 3.3 describes the hypothesized relationship between independent and dependent variables in a regression equation format. The independent variables, or three classes of psychosocial determinants, individual characteristics, social circumstances, and environmental conditions, taken together predict pathological fire behaviors,

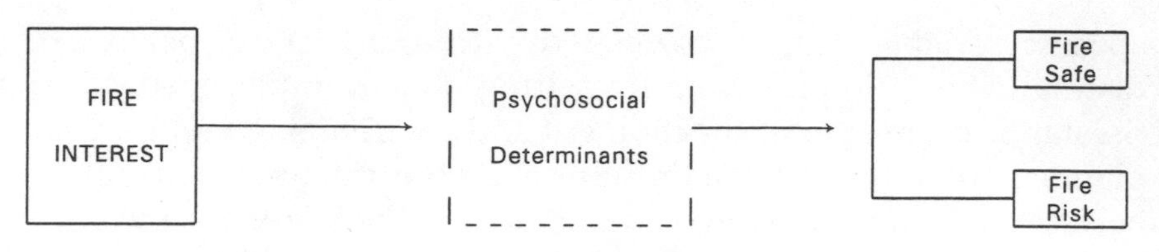

Figure 3.2. A predictive model of youthful fire behavior.

including fireplay or firesetting. Pathological fire behaviors will be accounted for by a combination of these classes of psychosocial determinants.

Measurement

A major property of a predictive model is its ability to be empirically evaluated. The independent variables, or three classes of psychosocial determinants, have been conceptualized into dimensions (see Table 3.1) and operationally defined into variable clusters (see Table 3.2). The dependent variables, or pathological fire behaviors, have been distinguished by behavioral correlates (see Table 3.3). Therefore, each of the components of the predictive model has been specified into quantifiable units of measurement. In addition, because this predictive model utilizes a regression format, additional relationships between classes of independent variables and between independent and dependent variables can be examined. For example, the relative contribution of each of the three classes of psychosocial determinants can be evaluated statistically by assigning different numerical weights or values to each of the three classes of independent variables. A variety of statistical relationships can be evaluated including those between variable clusters, dimensions,

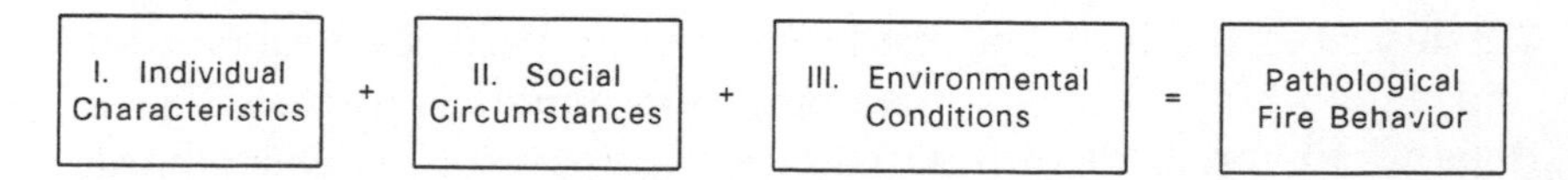

Figure 3.3. An hypothesized relationship between psychosocial determinants and pathological fire behavior.

and classes of psychosocial determinants, as well as between the determinants and the dependent variable of pathological fire behaviors.

Prediction

Another major property of a predictive model is that it generate hypotheses explaining or predicting the relationships between variables and observed behaviors. In particular, these hypotheses should deduce the nature of the relationships or make inferences on the directionality of the relationships. For example, it has been hypothesized that there is a specific age range associated with pathological fire behaviors. Previous research has verified a specific age range. Additional hypotheses now can be generated regarding chronological age and the severity of pathological fire behavior. It may be hypothesized that younger children are less likely to be involved in serious recurrent firesetting activity. Although it is premature to predict directionality on many of the independent variables hypothesized to be related to pathological fire behaviors, this predictive model allows for the hypothesis to be conceptualized and directionality to be empirically evaluated.

Application

The primary application of this predictive model for mapping the independent variables related to pathological fire behaviors is that it integrates the current body of knowledge in the field of youthful fire behavior. Second, the usefulness of the model should be tested in the quantity and quality of empirically based studies that it generates. Third, in addition to its ability to organize knowledge and generate research, this predictive model should be applied to the clinical description, diagnosis, and treatment of children involved in pathological fire behavior. It is the tripartite application of this predictive model which will help relieve the significant problem of youthful involvement in firesetting and reduce the high rate of juvenile-related arson fires.

SUMMARY

A predictive model is proposed integrating previous theories and research into a system for predicting the occurrence of youthful fire behavior. The underlying assumption of this model is that youthful fire behavior is a part of the normal developmental process of childhood. This behavior is expressed in terms of fire interest and occurs in most children as early as age three. Fire interest can lead to fire-safe or fire-risk behaviors. Children who are fire-safe have learned to use firestarting materials in a supervised setting and can demonstrate fire competence. Fire-risk behaviors, fireplay and firesetting, are predicted by specific psychosocial determinants. There are three major classes of psychosocial determinants—individual characteristics, social circumstances, and environmental conditions—which predict the pathological fire behaviors of fireplay and firesetting. This predictive model has three major applications. First, it provides an integrated framework for organizing the current body of knowledge in the field. Second, it generates hypotheses that can be empirically evaluated regarding the prediction of youthful fire behavior. Third, the knowledge accrued within the framework can be applied to the clinical description, diagnosis, and treatment of youngsters involved in pathological firesetting.

4

Pathological Fire Behavior—The Clinical Picture

An effective, integrated framework for organizing knowledge in the field of youthful fire behavior should have a major application in the clinical description and diagnosis of youngsters exhibiting pathological fire behavior. Although research may proceed in the field, it is of immediate importance for professionals to recognize patterns of firesetting, determine their severity, establish a clinical diagnosis, and eventually recommend one or more intervention strategies. Pioneering work is described which represents the first attempt at identifying youngsters who are at risk for becoming involved in patterns of serious, recurrent firesetting. These youngsters are defined as being at "clinical risk," and it is suggested that their firesetting behavior needs immediate professional attention. A method is presented which outlines the criteria for determining how observed firesetting behavior is categorized as significant psychopathology. The critical elements for identifying the specific features associated with the psychopathology of firesetting are presented as a way to organize and evaluate clinical information on youngsters. A system for arriving at a clinical diagnosis is described utilizing the current classes of major childhood psychiatric disturbances.

Finally, consideration is given to the question of whether pathological firesetting is a symptom of an existing class of psychiatric disorders or whether patterns of recurrent firesetting represent a new, major psychiatric syndrome.

YOUNGSTERS AT RISK

Youngsters are at risk for becoming involved in pathological firesetting when their normal interest in fire and subsequent developmental pattern of learning fire-safe behaviors becomes interrupted by particular psychosocial factors. When one or more of a specific set of psychosocial determinants exists within the lives of youngsters and their families, the probability increases that pathological fire behavior will occur. The question becomes, given certain psychosocial conditions, how can firesetting risk be determined for specific types of children and their families?

Previous pioneering work, developed for use by fire service personnel, describes how to interview and evaluate youthful firesetters ranging in age from 7 to 14 (FEMA, 1979, 1983). This system represents the first attempt at classifying pathological fire behavior for frontline personnel, like fire department arson investigators, who first come into contact with youngsters lighting fires. This method will be outlined briefly because it lays the foundation for assessing whether firesetting incidents represent childhood curiosity about fire or are indicative of more serious, underlying psychological problems. Subsequent to describing this system for assessing a youngster's clinical risk, a framework will be proposed for evaluating the psychological profiles of children and families who present with problems of pathological fire behavior. This framework utilizes three critical psychosocial elements—individual characteristics, social circumstances, and environmental conditions—as the major factors contributing to the emergence of pathological firesetting. These critical elements serve as a guide for assessing the nature and degree of individual and family pathology related to the problem of firesetting behavior. Once these critical psychosocial elements are described, the specific individual and family factors are identified which lead to a clinical description and diagnosis of youngsters involved in pathological firesetting.

Clinical Risk

Previous work describing childhood firesetting focused on identifying whether the firesetting behavior was accidental and motivated by curiosity or whether it represented more serious underlying psychological problems (FEMA, 1979, 1983). The major purpose for developing a method of classifying firesetting behavior was to be able to teach frontline personnel, such as firefighters and arson investigators, who are often the first to reach children involved in firestarts, how to recognize and understand the nature and motivation of firesetting behavior. If frontline personnel can identify those youngsters whose firesetting is recurrent and represents underlying psychological problems, then they can refer these youngsters and their families to the appropriate helping professional or agency. Hence, through structured interviews with youngsters and their families, frontline personnel can classify the nature and severity of detected firesetting behavior.

This classification system identifies three primary levels of clinical risk (FEMA, 1979, 1983). They are little, definite, and extreme, each representing successively more severe forms of firesetting behavior. By analyzing the information obtained in individual and family interviews and a written questionnaire, a numerical rating system determines whether youngsters can be classified as at little, definite, or extreme risk for firesetting. This classification system emphasizes the importance of the home and family, peers, school, and the specific firesetting behavior in determining risk. Youngsters classified as little risk are well adjusted interpersonally and socially, but have become involved in one or two firestarts because of curiosity and exploration or by accident. Definite-risk youngsters exhibit recurrent firesetting behaviors, where numerous firestarts represent the expression of underlying personal and social conflicts. Children classified as extreme risk also suffer from significant emotional and behavioral difficulties; however, for these youngsters firesetting already is a part of a firmly entrenched personality or behavior disturbance. This classification system suggests that the firestarting behavior of little-risk youngsters can be successfully remediated through educational intervention, and that definite- and extreme-risk youngsters and their families are at "clinical risk" and must be referred for professional help.

This firesetting behavior classification system provides frontline personnel with important triage skills for evaluating the nature and severity of firestarts initiated by children. In addition, the system puts into place a preliminary screening mechanism for the early identification of children and families who are at clinical risk and in need of specific professional help. Early identification of youngsters and families at clinical risk is the first step toward accurately describing the seriousness of firestarting behavior, detecting accompanying emotional or behavioral disorders, ascertaining a clinical diagnosis, and recommending appropriate intervention strategies.

Firesetting as Psychopathology

According to the classification system identifying little-, definite-, and extreme-risk categories of fire behavior, a single-episode firestart is most often initiated out of curiosity, exploration, or accident and can be successfully remediated through educational intervention. What happens when a firestart goes beyond curiosity or accident? What are the characteristics of recurrent, multiple firestarts? How can firesetting behavior be defined as a presenting clinical problem? These questions are critical to describing and evaluating firesetting behavior as clinical psychopathology.

Firestarting behavior must exhibit certain characteristics before it can be classified as psychopathology. Table 4.1 presents six factors that must be analyzed before firestarting can be categorized as a clinical disturbance. The major factors to be considered are history of firestarting, method of ignition, intention, ignition source, target, and behavior immediately after the ignition. For firestarting to be an indicator of psychopathology, at least four of the six factors, one of which must include a repetitive and persistent history, must characterize firesetting behavior.

Table 4.1
Factors Characterizing Firestarting as Psychopathology

Factor	Psychopathological Firestarting
I. History	Repetitive, persistent multiple firestarts for at least a six-month period
II. Method	Unsupervised, planned firestarts with specific targets of object or person
III. Ignition	Flammable or combustible materials actively sought or gathered for ignition
IV. Intention	Nonproductive firestarts for the purpose of malicious mischief, watching fire burn, revenge, profit, or harming person, property, or thing
V. Target	Target typically is other than self and is most often another person's property
VI. Behavior	There will be no voluntary admission of firesetting and immediately following ignition an attempt is made to watch the fire burn

All six of the factors characterizing firestarting as psychopathology are important; however, the most essential feature is the history of the firesetting behavior. Isolated, single episodes of firestarting do not necessarily represent significant behavioral disorders. For firestarting to be considered a significant behavioral disturbance, there must be a history of multiple firestarts taking place for at least a six-month period of time. The nature and extent of these firestarts may vary; they can range from parents finding spent matches in their youngsters' room to the burning pot on the stove. Most typically, these recurrent firestarts result in no serious fire damage, but their repetitive and persistent nature is the indication of a distinguishable behavior pattern.

The method of firestarting is critical in evaluating the severity of the firesetting behavior. Recurrent firestarting usually is a planned event as opposed to a spur-of-the-moment, impulsive activity. Matches and lighters are searched for and acquired. The firestart will take place in a concealed or isolated area where there is little possibility of immediate detection by an adult or authority figure. Often there is a specific target or place for setting the fire. Most all of the events leading up to the actual ignition are the result of a plan as opposed to a sudden impulse or accident.

In addition to the methods leading up to the firestart, the nature of the ignition is an important feature. The primary characteristic of pathological firesetting is that there is an active seeking out of ignition sources as well as flammable and combustible materials. Matches and lighters are searched for, acquired, and often hidden or concealed until they are needed. When a fire is set, there is an attempt to gather flammable materials, such as old newspapers, rags, or chemicals such as paints or alcohol to use as aids in spreading the fire. The physical conditions of where the fire is set also are considered important. That is why many firestarts begin in dried brush, wildland, or forests. The characteristics of the ignition demonstrate the sophistication of the firesetting behavior.

Perhaps one of the most clinically informative features of pathological firesetting is the intention, motivation, or reason behind the behavior. Most all repetitive firestarts are nonproductive in that there is no useful purpose for the fire. In fact, there are five most frequently occurring reasons for nonproductive firestarts. First, the firesetting often is motivated by malicious mischief and includes accompanying antisocial behaviors such as burglary and vandalism. Second, fires frequently are started for the purpose of watching objects or property burn. Third, feelings of revenge and anger toward close friends, loved ones, or even society in general are expressed by igniting a fire. Fourth, there may be fees to be earned or profits to be made by firesetting. And, fifth, there often is a specific, malicious intention to destroy property or to harm animals or persons. All of these motivations represent destructive, violent acts which run contrary to societal standards and which jeopardize the safety and well-being of others.

The target of the firestart often has significant meaning for the firesetter. It is important to pay close attention to what is set on fire. If personal possessions are ignited, the type of object and its potential significance to the firesetter should be considered. If another's property is ignited, as is usually the case, an analysis should be made of the relationship between the firesetter and the person whose property was set on fire. Although fires directed at animals, other persons, or the self are relatively infrequent, their occurrence indicates a far more serious behavioral disturbance than firestarts aimed at property.

Behaviors that occur immediately following firestarts indicate the degree of responsibility firesetters assign to their behavior. Youngsters involved in recurrent firesetting often watch the set fire burn. They may control the fire themselves and even participate in extinguishing it. However, they rarely voluntarily admit to their involvement in the ignition. If the fire is out of their control, rather than calling for help, they are more likely to run away to a safe spot. Often they will watch the fire burn and await the arrival of firefighters and emergency vehicles. There have been some anecdotal reports of firesetters actually volunteering to help with firefighting or other rescue activities associated with the fire. Firesetters often are most conflicted after the ignition, and their behavior reflects a confused need to both observe as well as suppress the evidence of their aggressive and violent act.

These six factors characterize acts of firestarting as psychopathology. All of these factors should be considered in evaluating firesetting as a significant clinical problem. The description of the actual firesetting incidents, utilizing these six factors as indicators of the nature and severity of existing psychopathology, will yield rich clinical information. Once an assessment of the pattern of firesetting has been established, the critical elements that combine to produce this type of behavior can be analyzed.

CRITICAL ELEMENTS IDENTIFYING PSYCHOPATHOLOGY

Once a determination has been made establishing the pattern of firesetting as an indication of psychopathology, it then becomes important to describe the critical elements that contribute to the behavioral disturbance. These critical elements—individual characteristics, social circumstances, and environmental conditions—have been identified previously within the context of presenting a conceptual model for predicting the occurrence of firesetting behavior (see Chapter 3). Their application within a clinical setting will introduce a systematic method for organizing clinical observation and information. When these three critical elements are clinically assessed, they will provide specific information to pinpoint the etiology of pathological firesetting as well as contribute to planning an effective intervention strategy to remediate the behavior.

The three critical elements of individual characteristics, social circumstances, and environmental conditions and their associated dimensions provide a conceptual structure for organizing and assessing clinical information related to pathological firesetting. There is a current body of evidence, consisting primarily of clinical studies and a small number of empirical investigations, which present a description of the psychodynamics of pathological firesetting behavior. This information will be reviewed within the conceptual framework of the three critical elements and their associated dimensions. An analysis of the current clinical and empirical information utilizing these critical elements as an organizing framework will yield a detailed clinical description of the psychosocial factors contributing to the emergence of pathological fire behavior. Table 4.2 summarizes the critical elements, their associated dimensions, and the clinical characteristics related to pathological firesetting.

Individual Characteristics

There are six major dimensions describing the individual characteristics of youngsters involved in pathological firesetting. They are demographic, physical, cognitive, emotion, motivation, and psychiatric dimensions. An analysis of these six dimensions reveals several major clinical features characteristic of young, recurrent firesetters.

Initial as well as more recent descriptive studies and empirical investigations demonstrate that males are more likely to be involved in 9 out of 10 recurrent firestarts (Lewis & Yarnell, 1951; FEMA, 1979, 1983; Fineman, 1980; FBI, 1982; Kolko et al., 1985). Clinical cases show that children as young as five years (Gaynor, 1985) and as old as 16 (Gruber et al., 1981; Strachan, 1981; Ritvo et al., 1982) have been involved in multiple firestarts. However, youngsters 10 years and older are most likely to exhibit recurrent firesetting behavior (Stewart & Culver, 1982). There is conflicting evidence with respect to the socioeconomic status of these young boys, with some studies reporting that childhood arson is a white, middle-class crime (Wooden & Berkey, 1984), others reporting a strong representation of the lower income groups (Strachan, 1981; Gruber et al., 1981; Heath et al., 1983), and still others finding no relationship between

Table 4.2
Clinical Characteristics Related to Pathological Firesetting

Critical Elements	Dimensions	Clinical Characteristics
I. Individual characteristics	A. Demographic	Young males, usually 10 years or older, from a mixed socioeconomic background
	B. Physical	Some evidence of organic deficiencies; a higher incidence of physical illnesses; enuresis; sexual abnormalities, including abuse; and high levels of energy
	C. Cognitive	Normal ranges of intellectual capabilities, with some evidence indicating the presence of learning disabilities
	D. Emotion	Tendency to experience overwhelming anger and inappropriately express aggression
	E. Motivation	Multiple intentions including the expression of displaced revenge, nonspecific anger, and the need for attention
	F. Psychiatric	Previous psychiatric history and diagnosis including conduct or personality disorder
II. Social circum stances	A. Family	Single-parent families where there is evidence of parental distancing and uninvolvement; high expression of aggressive behavior, including physical abuse; significant stressful family events such as numerous geographical relocations; and psychiatric histories for one or more parents
	B. Peers	Interpersonal maladjustment, including inabililty to initiate and maintain significant interpersonal relationships
	C. School	Below-average academic achievement with marked behavioral problems including suspension or expulsion from school
III. Environmental conditions	A. Antecedent stressors	Stressful life events of either a positive or negative nature including divorce, recent geographical move, or expulsion from school
	B. Behavioral expression	Expression of revenge or anger or need for attention from significant others
	C. Consequences	Little discipline by adults following initial firesetting incidents coupled with the lack of available professional help, including methods of retribution

socioeconomic status and pathological firesetting (Kuhnley et al., 1982; Kolko et al., 1985). Hence, the demographic profile of recurrent firesetters seems to be young males of 10 years or older who come from a variety of socioeconomic backgrounds.

There are a number of studies reporting significant physical characteristics and behaviors of pathological firesetters. Some investigators observed organic deficiencies in recurrent firesetters; however, these abnormalities were mostly chromosomal in nature and detected in adult populations (Hurley & Monahan, 1969; Nielsen, 1970). Some studies have reported high incidences of physical illnesses in young, recurrent firesetters, with a preponderance of allergies and respiratory problems (Siegelman, 1969; Siegelman & Folkman, 1971). There also is conflicting theory and evidence linking enuresis and abnormal sexual behaviors to pathological firesetting (Stekel, 1924; Lewis & Yarnell, 1951; Kaufman et al., 1961; Heath et al., 1976; Vreeland & Waller, 1979; Sakheim et al., 1985). More recent clinical observation is detecting a relationship between sexual abuse and recurrent firesetting (Gaynor, 1985). Finally, descriptive studies show that some youngsters exhibited hyperactive behavior, with typically short attention spans and excessively high levels of energy (Broling & Brotman, 1975; Kafry et al., 1981; Gruber et al., 1981). Although current empirical evidence is inconclusive concerning the strength of the relationship between specific physical characteristics and pathological firesetting, young, recurrent firestarters may exhibit one or more of a variety of physical characteristics, including organic deficiencies, a higher-than-normal incidence of somatic complaints, bed-wetting, sexual difficulties, and excessively high levels of energy.

The majority of studies evaluating the cognitive functioning of young firesetters show their capabilities to fall within the normal ranges of verbal and performance measures of intelligence (Benians, 1981; Kuhnley et al., 1982; Ritvo et al., 1982; Kolko et al., 1985). Although intelligence tests indicate normality, there is some evidence indicating that young, recurrent firesetters exhibit learning disabilities (Vandersall & Weiner, 1970; Kuhnley et al., 1982). Hence, although young, recurrent firesetters may be capable of normal intellectual functioning, their actual performance may be impeded by the existence of learning disabilities.

There is some clinical evidence suggesting that the underlying emotional state and ability to express experienced emotions may be related to recurrent pathological acts of firesetting (Block, Block, & Folkman, 1976; Awad & Harrison, 1976; Gaynor, 1985). Youngsters involved in repeated firestarts are primarily experiencing overwhelming feelings of aggression and anger. Because they lack both the ability to understand the reasons for their anger as well as the skill to appropriately express their feelings, they choose firesetting as a means of displaying their aggression. Although the link between multiple firestarts and emotional state and expression has been clinically observed, confirmation from empirically based investigations is necessary to verify this interpretation of recurrent firesetting behavior.

There is an apparent relationship between the emotional state and expression of youthful firesetters and the motivations underlying their behavior. Although there appear to be a number of different reasons for repeated firestarting, the common theme of these intentions seems to be that recurrent firesetting is an expression of internal conflict and emotional turmoil. The firestarts themselves may represent emotional discharges of displaced revenge or nonspecific anger (Lewis & Yarnell, 1951; Kaufman et al., 1961; Awad & Harrison, 1976; Gruber et al., 1981; Kolko et al., 1985). In addition, it has been suggested that firesetting indicates a need for recognition and attention with some clinical reports indicating recurrent firestarters have strong desires to be perceived as heroes (Lewis & Yarnell, 1951; Gaynor, 1985). There remains a great deal to be understood about the relationship between the underlying motivations of repeated firesetting and the overt expression of internal psychopathology.

Only recently have studies describing the clinical aspects of recurrent firestarters utilized formal psychiatric diagnostic procedures. There is some initial evidence indicating a relationship between pathological firesetting and the diagnostic category of conduct disorder (Kuhnley et al., 1982; Stewart & Culver, 1982; Kolko et al., 1985; Sakheim et al., 1985). Because firesetting is one of the descriptive behaviors utilized to assess the presence of conduct disorder, the relationship between this diagnostic category and pathological firesetting deserves further clarification. Clinical observation suggesting that repeated firestarts may represent the indirect expression of underlying aggression indicates a hypothesized relationship between firesetting and passive-aggressive personality disorder (Vreeland & Waller, 1979; Gaynor, 1985). There is no existing empirical evidence to support this hypothesis. In general, if youngsters have previous psychiatric histories, and if they have carried a diagnosis of conduct disorder, they may be at risk for becoming involved in pathological firesetting.

Clinical observation and empirical studies indicate that there are specific individual characteristics associated with recurrent firesetting. Although it is unlikely that youngsters presenting with pathological fire behavior will exhibit all of these clinical features, each of the six major dimensions should be evaluated. The probability is high that clinical assessment will reveal a significant number of individual characteristics consistent with the patterns of psychopathology indicated by current descriptive and empirical investigations.

Social Circumstances

There appears to be a set of social circumstances that influence the emergence of pathological firesetting. An analysis of this critical element can be divided into three dimensions—family, peers, and school. Each of these three dimensions to some extent reflects the ability to establish and maintain significant interpersonal relationships and to adjust and behave appropriately in social situations. A description of these dimensions will yield important information regarding how patterns of social interaction contribute to the development of recurrent firesetting behavior.

Family structure, patterns of family interaction, significant events, and parental pathology are related to the emergence of pathological firesetting behavior in youngsters. There is clinical evidence showing that many young, recurrent firesetters come from either single-parent families or families where one of the two parents, typically the father, has been absent for long periods of time (Vandersall & Weiner, 1970; Fine & Louie, 1979; Gruber et al., 1981; Stewart & Culver, 1982). The patterns of interaction within the family have been characterized as ones in which mothers tend to be overprotective and fathers often are uninvolved but occasionally will administer overly harsh methods of discipline and punishment (Vandersall & Weiner, 1970; Siegelman & Folkman, 1971; Kafry et al., 1981). In addition, there is some evidence suggesting that parents of firesetting youngsters may exhibit more aggressive, violent behaviors within the family setting, which in turn influences the developing behavior patterns of their children (Patterson, 1978). This pattern of violent interaction within the family is represented by reports of young firesetters who have been physically abused by one or more parents (Gruber et al., 1981). Significant stressful events occurring within the family, such as numerous geographical relocations, have been shown to be related to recurrent firesetting behavior (Siegelman & Folkman, 1971). Finally, although the evidence is conflicting, there have been studies showing that the parents of young pathological firesetters themselves have psychiatric histories, including diagnosis of antisocial personality and alcoholism (Fine & Louie, 1979; Stewart & Culver, 1982). These characteristics contribute to a dysfunctional family system that is conducive to the development of patterns of recurrent firesetting in youngsters.

Although patterns of interpersonal interaction are likely to originate within the family, the quality of peer relationships is often indicative of psychopathology. There is evidence suggesting that youngsters involved in recurrent firesetting are unable to initiate and maintain meaningful personal relationships and, as a consequence, experience feelings of exclusion and loneliness (Vandersall & Weiner, 1970; Heath et al., 1983). This social isolation may be related to the observation that these youngsters have unusually strong needs for attention motivating them to engage in dramatic behaviors which are likely to result in the desired attention, namely firesetting. One possible option for these youngsters is social skills training which will teach them how to establish and sustain healthy and significant friendships so that they need not engage in behaviors that are designed solely for dramatic effect and negative attention. In addition, social skills training may help these youngsters learn to participate with peers in mutually agreed-upon and socially acceptable forms of recreational activities.

The third dimension of social circumstances is school performance. Studies indicate that, despite normal intellectual capabilities, recurrent firesetters experience marked academic and behavior difficulties within the school setting (Vandersall & Weiner, 1970; Gruber et al., 1981; Kuhnley et al., 1982). Academic underachievement is a predominant problem for these youngsters, with many of them receiving unsatisfactory grades and falling one or more school years behind their chronological ages. School disruption also is typical, with these youngsters often involved in fights with peers and having histories of suspension or expulsion from school because of behavioral problems. Hence, recurrent firesetters have significant problems adjusting to school environments, which further adds to their social isolation and feelings of frustration and anger.

An unhealthy family atmosphere, unsuccessful peer relationships, and poor school adjustment comprise the social circumstances that contribute to the emergence of recurrent firesetting behavior in youngsters. Although these factors also may characterize other childhood behavior disorders, when they are considered along with the existence of specific individual characteristics and immediate environmental circumstances, the probability increases that patterns of firesetting will be observed. Clinical evaluation of habitual firesetters is likely to reveal the described pathology in all of the three dimensions of social circumstances. The social environment of these youngsters is the critical element which sets the stage for youngsters with predisposing individual characteristics to become involved in pathological firesetting.

Environmental Conditions

The critical element of environmental conditions is defined as the set of contingencies occurring within the immediately surrounding environment which elicit and reinforce patterns of firesetting behavior. Antecedent stressors, behavioral expression, and consequences comprise the three dimensions of environmental conditions. These dimensions alone cannot elicit firesetting, but rather they must be considered within the context of predisposing individual characteristics and specific social circumstances. If the critical elements of individual characteristics and social circumstances are indicative of existing psychopathology, then what begins as single-episode firestarts is likely to develop into patterns of recurrent firesetting behavior as a result of the influence of environmental conditions.

Antecedent stressors are those events which occur immediately prior to a firestart (Fineman, 1980). The specific time frame may range from a few minutes to six months prior to the firesetting. Typically, antecedent events are discrete occurrences which are of a stressful nature to youngsters who already are having difficulties adjusting to their life circumstances. There are a number of stressful events which could trigger firestarts such as divorce, recent geographical move, or expulsion from school. During clinical evaluation it is important to assess whether any significant changes have recently occurred within the lives of these youngsters, including family events, interpersonal relationships, or social circumstances. Although it may take some careful reconstruction of recent past events, it is likely that such an analysis will reveal one or more stressful life events related to a firestart. There is a growing body of evidence suggesting that the firestart itself may represent some type of behavioral expression (Awad & Harrison, 1976; Fineman, 1980; Gaynor, 1985). In addition, some work suggests that recurrent firesetters as a group tend to participate in more externalizing or acting-out behaviors as opposed to internalizing or problem-solving behaviors (Heath et al., 1983; Kolko et al., 1985). If this is the case, then it is crucial to understand what meaning these youngsters attribute to their own firesetting behavior. Considering previous interpretations, it is likely that the firestart itself may represent an expression of aggression or the need to be recognized by significant others or the community. Also, if firesetting can be interpreted as a pattern of behavioral expression, then the very act of starting a fire becomes reinforcing in and of itself. Hence, the single-episode firestart quickly develops into a serious pattern of recurrent firesetting, and it becomes increasingly more difficult to break the chain of mutually reinforcing events.

The specific conditions following firestarts which impact recurrent firesetters have received little attention. Kolko (1983) suggests there are four major areas which define the potential consequences of firesetting activities. They are medical (death, injury), legal (arrests, convictions), social (parental separation, removal from home), and financial (property loss, retribution). A fifth consequence can be added to the list, namely intervention or treatment programs initiated subsequent to the firesetting behavior. Kolko believes that an examination of these consequences will reveal information about the severity of firesetting. What may be discovered is a remarkable lack or absence of relevant consequences following a firestart. Not only may recurrent firesetters not understand the potential consequences of their behavior, but their firesetting may not have resulted in the appropriate implementation of these consequences. As a result, an implicit message is given that firesetting does not have personal consequences and therefore is an activity that reinforces anonymous participation.

Antecedent stressors, behavioral expression, and consequences are suggested to play a significant role in eliciting and reinforcing multiple firestarts. These immediate environmental conditions appear powerful in their ability to sustain patterns of firesetting. However, these patterns are initiated only when predisposing individual characteristics and particular social circumstances indicate the existence of significant psychopathology. The specific methods by which these environmental conditions influence firesetting is perhaps the most uncharted area in terms of verifying what has been hypothesized to occur. Clinical observation and empirical investigations should be guided by some of the suspected environmental conditions impacting recurrent firesetting, but they also should be open to uncovering new conditions or substantiating whether the current ones are in fact playing a significant part in maintaining pathological fire behavior.

A Composite of Critical Elements

The three critical elements of individual characteristics, social circumstances, and environmental conditions provide a composite profile of youngsters involved in pathological firesetting. First, the firesetting behavior itself is identified as pathological utilizing the criteria outlined in Table 4.1. Second, by defining the particular characteristics of the three critical elements and their associated dimensions, the specific psychopathology underlying the firesetting behavior is identified. The result is a detailed clinical description of youngsters presenting with patterns of recurrent firesetting behavior.

The exact mechanism of how these three critical elements of individual characteristics, social circumstances, and environmental conditions interact to produce patterns of pathological firesetting is not yet understood. Nor is it clear why the described psychopathology leads to firesetting behavior as opposed to other antisocial, acting-out behaviors. It appears as if current clinical and empirical information describing the psychopathology of firesetting raises more questions than it answers. However, given what is known about the three critical elements and their associated dimensions, can we arrive at an adequate clinical diagnosis to describe the observed behavior? Furthermore, does the pattern of childhood firesetting behavior represent a symptom of a larger behavioral disorder complex, or can it be identified as a unique psychiatric syndrome? These questions are the focus of the remainder of this chapter describing the clinical picture of pathological fire behavior.

TOWARD A CLINICAL DIAGNOSIS

Given what is known about the psychopathology of childhood firesetting, the next step is to designate those clinical diagnostic categories which best describe the behavior of these youngsters. Utilizing the guidelines described by the *Diagnostic and Statistical Manual, Third Edition* (DSM-III) (APA, 1980), the major classes of childhood psychiatric disturbances will be reviewed briefly to rule out those disorders which clearly are not reflective of recurrent firesetting behavior. The specific categories of information utilized by the DSM-III will be applied to the identified psychopathology of firesetting to determine the relevant diagnostic categories. It is suggested that there is a group of clinical diagnoses which describe youngsters involved in recurrent firesetting. However, it also is noted that youthful firesetting perhaps can be regarded as more than a single symptom or behavioral correlate of the existing system of psychiatric diagnostic categories.

Classes of Childhood Disorders

According to the DSM-III, there are five major classes of psychiatric disorders usually first evident during childhood, including infancy and adolescence. They are intellectual, behavioral, emotional, physical, and developmental. In addition, affective disorders and schizophrenia are psychiatric disturbances that can be diagnosed during childhood as well as adulthood utilizing the same criteria. These seven categories define the current universe of psychiatric disorders which may be applied to youngsters exhibiting firesetting behavior.

The first step in arriving at a clinical diagnosis is ruling out categories that do not adequately describe the primary psychopathology of recurrent firesetters. The major intellectual diagnostic category of childhood is mental retardation. The essential features characterizing this diagnosis do not exist among the population of recurrent firesetters. This is not to say that youngsters diagnosed as mentally retarded will never exhibit firesetting, rather mental retardation is unlikely to be the primary diagnosis for the majority of youngsters involved in recurrent firesetting.

Other major classes of psychiatric disturbances which can be ruled out are physical and developmental disorders. The class of physical disturbances includes eating and movement disorders and other disorders with physical manifestation. Although there is one reported case in the literature of a youngster diagnosed as developmental disabled becoming involved in recurrent firesetting (Kolko, 1983), the relationship between this diagnostic category and pathological firesetting is infrequent. Therefore, it appears as if the physical and developmental classes of psychiatric disturbances are likely to be ruled out as major indicators of recurrent firesetting.

There are some studies showing a relationship between youngsters diagnosed as psychotic and pathological firesetting (Kaufman et al., 1961; Vandersall & Weiner, 1970). There also is some work indicating a relationship between childhood schizophrenia and firesetting (Bender, 1959). Hence, the major classes of psychotic disturbances must be considered when diagnosing children presenting with pathological firesetting. The essential features of both affective disorders, including manic or depressive episodes, and schizophrenia, including characteristic symptoms such as disturbed thought content, altered perceptions, inappropriate affect, and withdrawal, must be evaluated. The current clinical incidence rate of detecting major psychotic disturbances in youthful firesetters ranges from 5% to 33% with the more frequently occurring incidence rate being at the 5% to 10% level (Gaynor, 1985). Given the implications of diagnosing childhood psychosis, these major classes of psychiatric disturbances must be considered, the essential features of these disorders analyzed, and the relationship between the presenting problem of pathological firesetting and a potential psychotic diagnosis specified. If the essential features of affective disorders and schizophrenia are not present, then the probability of pathological firesetting reflecting a psychotic disturbance will be low. However, if one or more essential features are detected, then the diagnosis of psychosis becomes more likely and should be seriously considered as a potential diagnostic category.

The remaining classes of childhood psychiatric disturbances, emotional and behavioral disorders, are highly probable diagnostic categories to be considered when the presenting problem is pathological firesetting. The emotional disturbances include anxiety, identity, schizoid, and oppositional disorders. The behavioral disturbances include attention deficit and conduct disorders. The current clinical and empirical evidence suggests that these two classes of psychiatric disturbances—emotional and behavioral disorders—are the most frequently occurring diagnostic categories related to recurrent firesetting (Kuhnley et al., 1982; Stewart & Culver, 1982; Kolko et al., 1985; Sakheim et al., 1985).

Clinical Diagnosis

If major psychosis has been evaluated and ruled out, then the emotional and behavioral classes of psychiatric disturbances must be assessed and carefully considered as the potential clinical diagnosis

The standard types of information must be evaluated when determining the diagnosis for the presenting problem of pathological firesetting. This information includes essential features, behavioral problems, age at onset, predisposing factors, prevalence, sex ratio, and familial patterns. Each of these types of information must be carefully assessed to arrive at a clinical diagnosis.

The class of emotional disturbances includes anxiety, identity, schizoid, and oppositional disorders. The essential factors describing anxiety and identity disorders do not fit the identified psychopathology associated with recurrent firesetting. However, some of the essential factors defining the Schizoid Disorder may apply to recurrent firesetters. For example, one feature of youngsters diagnosed as Schizoid is their inability to form social relationships with anyone of a similar age. Consequently, they display a marked degree of social isolation. In addition, these children display occasional outbursts of aggressive behavior. The disorder always begins in childhood and is much more common in boys than girls. The primary differential diagnosis is Conduct Disorder. Hence, if psychosis is not evident, a Conduct Disorder diagnosis is ruled out, and a marked degree of social isolation is the predominant feature accompanying the recurrent firesetting, then Schizoid Disorder can be considered as a clinical diagnosis.

The emotional disturbance of Oppositional Disorder also is related to psychopathological firesetting. The essential feature of this disorder is a pattern of disobedient, negativistic, and provocative opposition to authority figures, including parents and teachers. Behavioral opposition may take the form of violating minor rules and displaying temper tantrums or extreme argumentativeness. The disorder surfaces during childhood, is chronic in nature, and usually results in poor social relationships, school failure, and involvement in substance abuse. The primary differential diagnosis is Conduct Disorder. However, conduct disorders are differentiated from oppositional disorders in that youngsters must violate the basic rights of others or major age-appropriate societal norms to be classified as a Conduct Disorder diagnosis. In addition, what may start out as an Oppositional Disorder may later turn out to be an early manifestation of a Conduct Disorder. Hence, those youngsters in the beginning phases of recurrent firesetting, provided their individual characteristics and social circumstances are consistent with the diagnosis of Oppositional Disorder, initially may be so classified. If the recurrent firesetting behavior patterns continue, however, the diagnosis is likely to shift to Conduct Disorder.

The remaining major psychiatric disturbance indicative of pathological firesetting is behavioral disorders. Both attention deficit and conduct disorders have been recognized as the most frequently occurring diagnostic categories describing youngsters exhibiting recurrent firesetting behavior (Siegelman & Folkman, 1971; Kuhnley et al., 1982; Stewart & Culver, 1982; Heath et al., 1983; Kolko et al., 1985; Sakheim et al., 1985). The major diagnostic trend is to classify youngsters involved in pathological firesetting primarily under conduct disorder, with the additional diagnosis of attention deficit disorder less frequently warranted.

There are two types of attention deficit disorders, one which is accompanied by hyperactivity and the other which is without hyperactivity. The essential features of both are developmentally inappropriate inattention and impulsivity, usually first noticed within the school setting. Academic difficulty is common, and social and interpersonal adjustment is frequently poor. If there is accompanying hyperactivity, there may be an excessive amount of haphazard, poorly organized gross motor activity. Youngsters involved in recurrent firesetting usually display many of the features descriptive of Attention Deficit Disorder with the additional, but predominant behavior of firestarting.

Perhaps the most accurate discription of identified psychopathology associated with recurrent firesetters can be found in the clinical diagnosis of Conduct Disorder. The essential feature of the disorder is a repetitive and persistent pattern of behavior that either violates the basic rights of others or breaks major age-appropriate societal norms. There are four specific subtypes of the disorder based on the presence or absence of social bonds and the presence or absence of a pattern of aggressive antisocial behavior. Pathological firesetting is most closely related to two subtypes, both of which are characterized by aggressive behavior.

The first subtype, Conduct Disorder, Undersocialized, Aggressive, is characterized by physical violence against property or thefts outside the home and failure to establish a normal degree of affection, empathy, or bond with others. The second subtype, Conduct Disorder, Socialized, Aggressive, includes the same patterns of violent, antisocial behavior but there is evidence of social attachment to others. Both of these subtypes also are characterized by the following features. Youngsters feel unfairly treated and their self-esteem is low. Academic achievement is below the level expected on the basis of intelligence and age. Behavioral problems include school suspension, legal difficulties, and excessive physical injury resulting from accidents and fights with peers. The age of onset for these conduct disorders is prepubertal for the undersocialized subtype and pubertal for the Socialized subtype. Some predisposing factors associated with the two subtypes are parental rejection, inconsistent discipline with harsh punishment, and frequent shifting of parental figures for the Unsocialized subtype and large family size with an absent father or a father with Alcohol Dependence for the socialized subtype. Both of the Conduct Disorder subtypes are common and occur more frequently among boys, with ratios ranging from 4 : 1 to 12 : 1. In addition, both subtypes are typically observed in families where one or more adults have previous psychiatric disorders of Antisocial Personality or Alcohol Dependence. Most all of the features associated with Undersocialized and Socialized Aggressive Conduct Disorder, including behavioral problems, age at onset, predisposing factors, prevalence, sex ratio, and familial pattern, fit the individual, social, and environmental descriptions of the psychopathology associated with youthful firesetting. Hence, the predominant clinical diagnosis for the majority of youngsters exhibiting pathological firesetting is Aggressive Conduct Disorder of either the undersocialized or socialized subtype.

The major classes of psychiatric disturbances and associated diagnoses that must be considered during a clinical evaluation of youngsters presenting with pathological firesetting are the psychoses of Affective Disorders and Schizophrenia, the emotional disturbance of Schizoid Disorder and Oppositional Disorder, and the behavioral disturbances of Attention Deficit Disorder and Aggressive Conduct

Disorder. The best method for arriving at a clinical diagnosis is first ruling out the possibility of psychosis and then proceeding to differentiate between emotional and behavioral disturbances. There is mounting clinical and empirical evidence suggesting that the diagnostic category which most frequently defines the associated psychopathology of recurrent firesetting is the Undersocialized or Socialized Aggressive Conduct Disorder.

Pathological Firesetting—Symptom or Syndrome?

Under the current system of describing and diagnosing pathological firesetting, firestarting is viewed as one symptom or behavioral correlate of a major psychiatric disturbance, which is indicated most frequently by behavioral disorders and the specific diagnosis of Aggressive Conduct Disorder. The question is raised as to whether pathological firesetting itself represents a distinguishable childhood syndrome and therefore should be recognized by a separate diagnostic category.

One method of addressing this question is to examine whether the specific features representing the psychopathology of recurrent firesetting are evident in those factors characterizing the psychiatric diagnosis of Conduct Disorder and Attention Deficit Disorder. Table 4.3 shows whether the critical elements describing the psychopathology of recurrent firesetting are evident in the factors characterizing the diagnosis of Conduct Disorder and Attention Deficit Disorder indicative of psychiatric behavioral disturbances.

An analysis of the information contained in Table 4.3 suggests that, overall, there is a moderate degree of consistency between the psychopathology identified with recurrent firesetting and the combined psychiatric diagnosis of Conduct Disorder and Attention Deficit Disorder. Of the 26 critical elements describing the psychopathology of recurrent firesetting, 15, or 58%, are evident in the two diagnoses representing psychiatric behavioral disorders. A more conservative estimate of consistency (adjusting for the two variables of enuresis and sexual abnormalities for which there is conflicting clinical and empirical support) indicates that 63% of the critical

Table 4.3
Comparing the Fit Between Firesetting Psychopathology and Psychiatric Diagnosis

Psychopathology Describing Recurrent Firesetting — Critical Elements Present	Psychiatric Diagnosis (Conduct Disorder/ Attention Deficit Disorder) Evident	
	Yes	No
I. Individual characteristics		
A. Demographic		
Young males	X	
Ten years or older	X	
B. Physical		
Organic deficiencies		X
Higher incidence of physical illness		X
(Enuresis)		(X)
(Sexual abnormalities)		(X)
High levels of energy	X	
C. Cognitive		
Normal range of intellectual capabilities	X	
Evidence of learning disabilities		X
D. Emotion		
Anger	X	
Expressed aggression	X	
E. Motivation		
Revenge	X	
II. Social circumstances		
A. Family		
Single-parent	X	
Distancing and uninvolvement	X	
Expression of aggressive behavior	X	
Stressful family events		X
Psychiatric history of parents	X	
B. Peers		
Inability to initiate and maintain significant interpersonal relationships	X	
C. School		
Below-average achievement	X	
Significant behavioral problems	X	
III. Environmental conditions		
A. Antecedent stressors		
Stressful life events		X
B. Behavioral expression		
Revenge/anger		X
Need for attention		X
C. Consequences		
Little discipline by adults or authority figures	X	
No professional help		X
No retribution		X

Critical elements included in parentheses indicate that current evidence is conflicting regarding their relationship to pathological firesetting.

elements describing the clinical features of pathological firesetting are evident in the class of psychiatric behavioral disturbances. If each of the three types of critical elements are analyzed independently, their degrees of consistency will vary. The degree of consistency for individual characteristics (adjusting for the variables of enuresis and sexual abnormalities) ranges from a moderate 58% to a relatively high 70%. For social circumstances, there is a relatively high degree of 88% consistency. However, for environmental conditions, there is a relatively low degree of 18% consistency. It must be noted that of the three types of critical elements, environmental conditions represent only those factors in the immediate environment which reinforce an already predisposed pattern of firesetting behavior. In other words, it is only one of the three elements (and perhaps the one that deserves the least weight) critical to the description of pathological firesetting. Hence, although there is a moderate degree of consistency for all three types of critical elements contributing to the composite description of the psychopathology of firesetting and the major class of psychiatric behavioral disturbances, the range of consistency varies from low to high when each of the three critical elements is independently considered. It may be artificial to evaluate each of the three critical elements separately because it is proposed that all of the critical elements must be considered as a composite description of the psychopathology indicative of recurrent firesetting.

The relative moderate degree of consistency between the critical elements and psychiatric diagnostic categories suggests that it may be premature to ascertain whether pathological firesetting is a single symptom or behavior reflective of the class of psychiatric behavioral disorders or whether it deserves recognition as a distinguishable psychiatric syndrome. Certainly, there is a significant amount of psychopathology associated with recurrent firesetting to raise the question of how to classify the pattern of psychiatric disturbance. For the present, it may be sufficient to propose that pathological firesetting is more than just a symptom of the diagnostic categories of Conduct Disorder and Attention Deficit Disorder and less than a new distinguishable psychiatric syndrome deserving of its own diagnostic classification.

SUMMARY

Criteria are presented for determining how patterns of firesetting are classified as a major clinical disturbance. These criteria include actual features of the firesetting behavior such as the history of firestarting, the method of setting fires, the ignition source, the intention of firestarts, the target of fires, and the behavior occurring immediately subsequent to the firestarts. If the presenting firesetting pattern meets these criteria, then the behavior is indicative of existing psychopathology. A system for organizing and evaluating the psychopathology of firesetting is reviewed which utilizes three major types of critical elements—individual characteristics, social circumstances, and environmental conditions—to present the current clinical and empirical description of pathological fire behavior. An approach for arriving at a clinical diagnosis is suggested which applies the current major classes of childhood psychiatric disturbances to the identification of those diagnostic categories most typically associated with recurrent firesetting. These psychiatric categories include the less frequently occurring psychotic diagnoses of Affective Disorders and Schizophrenia and the more frequently diagnosed emotional disturbances of Schizoid Disorder and Oppositional Disorder and behavioral disturbances of Attention Deficit Disorder and Aggressive Conduct Disorder. The clinical features associated with the psychiatric diagnosis of Aggressive Conduct Disorder demonstrate a moderate degree of consistency with the critical elements indicative of the psychopathology of youthful firesetting. It is concluded that pathological firesetting is more than a symptom of existing diagnostic categories but less than a distinguishable psychiatric syndrome deserving of its own diagnostic classification.

Section II
Intervention

5

Interviewing and Evaluating Youthful Firesetters

The clinical interview is one of the most essential methods for gathering information on youngsters presenting with pathological firesetting behavior. This interview should take place between a mental health professional, the youngster, and one or more parents or responsible adults. There are a series of general considerations to be observed which facilitate the process of the interview. The primary objectives of the clinical interview are to obtain data on those critical elements which both identify the existing psychopathology as well as describe the history and pattern of firesetting behavior, analyze this information to specify targeted behaviors associated with the observed psychopathology which must be changed, establish a working clinical diagnosis, and recommend an intervention strategy designed to eliminate firestarting and adjust the accompanying psychopathology. A comprehensive clinical assessment of the presenting problem of pathological firesetting and the related psychopathology is a necessary prerequisite to implementing an effective intervention strategy aimed at helping youngsters and their families remediate a serious and dangerous behavior pattern.

GENERAL CONSIDERATIONS FOR THE CLINICAL EVALUATION

Regardless of the presenting problem, clinical evaluations often share similar characteristics, such as the interview format, the preconceived attitudes of both the youngsters and their parents as well as those of the mental health professionals, and the style with which the interview is conducted. These features of the clinical interview are equally important in evaluating youngsters presenting with pathological firesetting behavior. In fact, clinical experience suggests that there are specific circumstances characterizing the firesetting problem itself which distinguish this type of clinical evaluation from other kinds of mental health assessments. Hence, what follows is a description of three general areas to be considered during the clinical evaluation—the interview format, preconceived attitudes, and interview style—and how these three areas relate particularly to youngsters presenting with the problem of pathological firesetting behavior.

The Interview Format

There are three important aspects of the clinical interview format to be considered when conducting an initial evaluation. They are who will be interviewed and how, the length of the interview, and the outcome or product of the clinical evaluation. Although these three interview format issues must be addressed by the individual clinician, there are some helpful guidelines that are likely to yield rich clinical data.

First, although youngsters present with the pathological firesetting problem and hence are the primary target of the clinical interview, their behavior does not occur as a single episode or symptom, nor does it occur without the influence of social circumstances and environmental conditions. Therefore, although it is of critical importance to interview youngsters individually concerning their firesetting behavior, it is also essential to evaluate the parents and their perspective on the problem. In addition, the direct observation of interaction between youngsters and parents often provides data on the sources of support or tension that exist within the family unit. It is recommended that interview time be given separately to evaluate both youngsters and their parents and that some consideration be given to a conjoint assessment of the family system.

Second, given that a clinical interview is suggested for both youngsters as well as their parents, it is likely that this type of evaluation will take a minimum of one hour. Clinical experience indicates that the maximum length of time for such an interview is three hours, with the majority of evaluations running about 90 minutes. At the beginning of the clinical evaluation, it is recommended that the entire family be interviewed and informed of the evaluation process and procedure. However, during the 90-minute period, the order of the remaining separate and conjoint interviews may vary depending on such considerations as the perceived comfort level of the youngsters and parents. For example, if a very young child appears anxious and concerned, the clinician may want to introduce some games or puppets to engage the youngster in conversation. If children are left alone waiting while their parents are interviewed, it is important that books, games, or some form of entertainment be available to them. Reading material on fire prevention is good to distribute. It is recommended that, primarily for the purpose of achieving closure, the entire family be together for the termination of the evaluation. Therefore, within an average 90-minute interview, it is possible to clinically evaluate youngsters and families presenting with pathological firesetting behavior.

Third, the outcome or product of the evaluation should be specified by the clinician to the family at some point during the interview. There are a range of possible outcomes, from verbal feedback and recommendations to written reports. Although the actual outcome of the evaluation often depends on its intended use, clinicians will have to decide on a case-by-case basis what they can offer. It is critical that whoever is requesting the evaluation understand and agree as to the result or product of the assessment.

A comprehensive evaluation of youngsters and their families within a 90-minute period which results in some type of tangible assessment is the key aspect of the clinical interview format. Clinicians must consider such issues as the order of evaluating youngsters and their families, the actual amount of time they will spend on the assessment, and the particular product they are willing to offer. The format of the clinical interview should facilitate the task of gathering clinical data while at the same time allowing youngsters and their families to feel comfortable telling their stories about experienced firesetting behaviors.

Preconceived Attitudes

Clinical experience suggests that, regardless of the presenting problem, both youngsters and their families as well as mental health professionals bring certain expectations with them to the clinical interview. This is a naturally occurring phenomenon, and it is most helpful to be aware of the orientation and content of these expectations. There are specific attitudes which frequently come to the surface during the clinical interview when the evaluation is focused on pathological firesetting. If there is a perception and awareness of these attitudes, then it is often therapeutic to discuss them openly during the clinical evaluation.

The most frequently occurring attitude that parents bring to the clinical interview is the expectation that the evaluation will serve as a form of retribution or punishment for the firesetting incidents. In addition, they hope their youngsters perceive the interview as the consequence that must be experienced as a result of their firesetting behavior. This attitude serves several useful functions for parents. It allows them to feel a sense of relief that they are finally able to find the help they need for their problem. In addition, it removes some of the responsibility from their shoulders because they now feel as if they are going to share in handling the consequences of their youngsters' firesetting behavior. It is important for clinicians to clarify the role and function of the evaluation so that parents can adjust their expectations and attitudes to a realistic level.

Youngsters entering the clinical interview often sense their parents' expectations of using the evaluation as some form of retribution or punishment. Hence, the most frequently occurring feeling children bring with them to the clinical interview is fear. They understand that their firesetting behavior has led them into this unknown situation, and they fear what they do not know or understand. Often, by simply acknowledging these fearful feelings youngsters become more relaxed and can be eased into participating in an effective interview.

If parents expect the clinical evaluation to serve as retribution and youngsters fear impending punishment, clinicians often experience feelings of being a substitute parent and an authority figure. There is a temptation to provide immediate structure and organization into a family system where it appears that none exists. Rescue fantasies are frequently experienced by clinicians. It is important to recognize these feelings and how they influence perceptions and behaviors during the interview. Although in many respects clinicians are the "experts" in the evaluation and treatment of pathological firesetting, the presenting behavior is best viewed as a problem to be resolved primarily by youngsters and their families under the guidance of mental health professionals.

It should be recognized that youngsters, parents, and clinicians all bring a set of preconceived attitudes with them into the clinical interview. These expectations are likely to influence the perceptions and behaviors of all those participating in the evaluation. The most predominant expectation is that the clinical assessment will somehow serve as the consequence or retribution of previous firesetting incidents. Hence, youngsters fear the evaluation, and clinicians' self-perceptions are of substitute parents and authority figures. It is important to acknowledge these and any other preconceived expectations so that communication channels are kept open to facilitate an effective clinical interview.

The Interview Style

Through experience, clinicians develop individual styles associated with conducting interviews and evaluations. There is a delicate balance between gathering the necessary clinical data and creating a comfortable interview environment for youngsters and their families. While clinicians must utilize their own effective interview style, these are a few techniques that facilitate the ability of youngsters and parents to talk about their experienced firesetting problem.

Frequently youngsters and their families are clinically evaluated immediately following a major firesetting incident. Consequently, parents are upset and angry and children are confused and frightened. The firesetting incident often is perceived as a crisis event. Therefore, the interview is likely to be stressful for the entire family. Clinical experience suggests that parents are eager to talk about the firesetting and associated problems but youngsters are more reluctant to share information. It is important for clinicians to let the parents tell their story. In general, the relevant clinical information will emerge with some gentle guidance by clinicians. However, during the evaluation the amount of clinical data offered by the parents also is directly related to how comfortable they feel telling their story.

In addition to listening attentively to parents, it is important to relate to the youngsters on whom most of the attention tends to be focused during the evaluation. The majority of youngsters, especially children between the ages of 6 and 10, will be very hesitant to talk about any topic within the clinical setting. Hence, when questions turn to their involvement in firesetting, there may be even more resistance to communicate. Often, initial interview time is well spent in letting youngsters adjust slowly to the idea of talking about their firesetting. Questions should be asked so that youngsters gain some understanding of themselves as they relate to the firesetting problem. For example, if it appears that a major fire was started as a result of an accidental, single-episode firestart, then it is appropriate to acknowledge this during the evaluation. Many youngsters need to have someone demonstrate that they understand and empathize with the motivations and circumstances underlying their firesetting behavior.

During the evaluation, clinicians must utilize their individual interview styles to listen attentively to parents and try to help youngsters understand themselves as they relate to their firesetting behavior. This must be accomplished within the context of gathering sufficient clinical data to identify the existing psychopathology, determine the severity of the firesetting behavior, specify targeted behaviors that must be changed, establish a clinical diagnosis, and recommend an effective intervention strategy to help youngsters and their families remediate the presenting problems.

THE CLINICAL DATA

The data collected during the clinical evaluation can be organized according to an interview plan. This interview plan contains two major classes of clinical data—those critical elements identifying existing psychopathology and those factors describing the firesetting behavior history. What follows are suggested methods for collecting clinical data relevant to specifying the existing psychopathology and detailing the history of firesetting behavior. Although individual styles and methods of data collection will vary, the information resulting from an interview plan must yield a precise clinical description of youngsters and their families. This clinical description will be utilized to identify behaviors targeted for change, formulate a working clinical diagnosis, and design an intervention plan to stop the firesetting behavior and adjust those underlying psychosocial determinants contributing to the presenting psychopathology.

Critical Elements Identifying Psychopathology

Previous chapters outlined the three types of critical elements—individual characteristics, social circumstances, and environmental conditions—which empirical studies and clinical observation suggest contribute to the psychopathology of firesetting. These three critical elements and their associated dimensions and variables must be evaluated during the clinical interview. Several different methods will be suggested for evaluating these critical elements within the context of the clinical interview. Clinicians must choose those methods which are most consistent with their style of evaluation and which will yield the most useful information about youngsters and families presenting with pathological firesetting.

There are five dimensions comprising the elements of individual characteristics which must be evaluated during the clinical interview. The first two dimensions, demographic and physical, can both be assessed by questioning. The demographic variables to be considered are the age, sex, and race of the youngsters as well as the socioeconomic background of their current living situation. The physical characteristics to be evaluated are general medical history, with a notation of any genetic or chronic physical diseases or long-term medication therapies; an analysis of sexual activity with attention paid to experiences of abuse; and an assessment of general energy level, focusing on the potential for a clinical diagnosis of hyperactivity.

The third individual characteristics dimension, cognitive, can be evaluated in a number of ways. The two variables comprising the cognitive dimension are intelligence level and learning ability. The majority of youngsters presenting with pathological firesetting behavior will have had some sort of school experience. Therefore, one method of informally assessing intelligence and learning ability is by questioning parents about progress and achievement within the school setting. There is a trend for these youngsters to demonstrate normal intelligence while exhibiting some patterns of learning disability. Consequently, their school performance is often below average. Hence, if there has not been an assessment of intelligence or learning ability and there is some evidence from school suggesting learning difficulties, then formal psychological testing may be indicated. If upon interviewing a youngster there is an observed verbal or performance deficit, some clinicians may elect to administer a brief assessment instrument for verification. A battery of psychological tests, focused on assessing intelligence and learning ability, is not typically included in an initial clinical evaluation of youngsters presenting with pathological firesetting, although it is frequently recommended as part of the intervention strategy.

The fourth dimension, emotion, is an individual characteristic which often emerges as one of the more salient features characterizing youngsters involved in pathological firesetting behavior. Clinical theory and observation suggest that these youngsters experience a predominant emotional state of anger. In addition, they are unable to recognize their feelings of anger and they tend to express them in socially unacceptable ways. Hence, the firesetting behavior of these youngsters can be interpreted as an expression of anger and aggression. There are a number of ways to evaluate the emotional adjustment of these youngsters. One method is simply by asking them what they are feeling most of the time and give them choices such as happy, sad, angry, frustrated, etc. A second method of assessing emotional state is to ask youngsters how they felt immediately before and after firestarting. A third, and perhaps more formal, method is by administering psychological tests such as the Children's Apperception Test (Buros, 1972) or the Little Bear Test (Block, Block, & Folkman, 1976). The latter assessment instrument has been adapted specifically to evaluate the emotional state associated with firesetting by the inclusion of two fire-related stimulus cards. Clinical assessment of emotion must focus on the predominant affective state of these youngsters, the emotions that accompany their firestarting, and their methods of expressing their feelings, especially anger and aggression.

The fifth individual characteristics dimension is psychiatric history. For younger children, under eight years of age, firesetting often is the first signal or evidence of underlying psychopathology. Therefore, there is not likely to be a previous psychiatric history. However, for youngsters eight years and older, firesetting is typically one of several exhibited pathological behaviors in need of professional attention. Hence, these youngsters often bring with them psychiatric histories consisting of outpatient and occasionally inpatient treatment. It is important to understand previous psychiatric evaluation and treatment episodes from the family perspective as well as from the responsible mental health professional. Adequate information regarding previous intervention will help put the current firesetting behavior in the appropriate clinical perspective.

The critical element of social circumstances consists of three dimensions—family, peers, and school. There is some clinical and research evidence suggesting that family characteristics may be the predominant factor responsible not only for the emergence of firesetting behavior, but also for its recurrence (Patterson, 1978; Fineman, 1980; Gaynor, 1985; Kolko et al., 1985). Consequently, it is crucial to conduct a thorough evaluation of the family to understand its influence on youngsters' behavior in general, and in particular, their involvement with firesetting.

There are four family features to be assessed during a clinical interview. They are structure, behavior, events, and pathology. Structure refers to the composition or members of the family and how they are related to one another. It is important to determine whether this is the biological family, and if not, whether any information can be obtained about the biological family. If youngsters are adopted, it is useful to know, if possible, the circumstances surrounding the adoption. The parenting structure must be assessed to determine whether both parents are living in the home and the various roles and responsibilities they assume for their youngsters. Finally, the sibling structure should be evaluated to determine the number, sex, and ages of brothers and sisters. In addition, the birth order of siblings is important because it may indicate special relationships between youngsters in the family. A description of the family structure provides clinical data on where and how youngsters fit within their primary interpersonal and social environment.

Family behavior refers to the way in which family members interact with one another. A clinical interview can assess the general family environment, the types of modeling that occur, and the methods of guidance, supervision, and discipline utilized within the family. An adequate evaluation of family environment can be achieved by asking members to describe their view of the family using various adjectives such as open, closed, permissive, authoritative, etc. A more formal method of assessing family environment can be conducted by administering the Family Environment Scale (Moos & Insel, 1974), although this measurement instrument is used more often for collecting research rather than clinical data.

In addition to describing the family environment, the type of behavioral modeling that occurs within the family must be evaluated. Because clinical theory suggests that firesetting youngsters may first learn this aggressive, violent behavior within the family system, it must be determined whether parents, siblings, or other significant role models are exhibiting firesetting or other aggressive, acting-out behaviors. These aggressive behaviors may occur between family members or may be initiated by family members to others outside of the family unit. All members of the family should be questioned about the occurrence and expression of aggressive, violent behavior.

Although it is important to identify significant role models and their influence on youngsters' behavior, the other types of interactions between family members also must be evaluated. The degree to which parents are involved in guiding and supervising their childrens' behavior will determine the quality of their relationship. Clinical theory indicates that parents of firesetters often are uninvolved with their youngsters. It also is reported that these parents may use overly harsh methods of punishing their youngsters. There have been a handful of case reports of firesetting youngsters being the victims of physical abuse (Gaynor, 1985). Not only must the negative or weak aspects of family interaction be assessed, but the positive, strong characteristics of family behavior must be identified. The emphasis should be on describing, from each family member's perspective, the quality of the relationship between family members.

The stressful events that occur within the family system can have a major impact on the behavior of youngsters. The effect of these events on behavior may be observed immediately after their occurrence as well as three to six months later. Therefore, family members should be asked to recall any significant family events occurring within the last year such as geographical moves, separation or divorce, additions of new family members, or deaths. The number, type, and frequency with which these events took place should be assessed. The occurrence of firesetting behavior is often linked both in time and in significance to the occurrence of stressful family events.

The final family feature that should be assessed during a clinical interview is whether individual members, especially the parents or responsible adults, have histories of or are currently exhibiting psychopathology. There is some clinical evidence indicating that parents of firesetting youngsters often have psychiatric histories which include diagnoses of antisocial personality and alcoholism (Fine & Louie, 1979; Stewart & Culver, 1982). Parental psychopathology can have a profound influence on the behavior of youngsters; therefore, it is critical to assess the psychiatric histories of all family members.

During the clinical interview there may be additional areas which emerge regarding the characteristics of families and their patterns of interaction. These areas of importance should be pursued by clinicians to determine their relevance to the observed firesetting behavior. Clinicians must arrive at an accurate description of the family dynamics, assess the reaction of family members to the occurrence of firesetting, and ascertain what adjustments the family will make to participate in the remediation of the firesetting behavior.

The remaining two dimensions of social circumstances—peer relationships and school performance and behavior—also are important areas for assessment. There is evidence suggesting that firesetting youngsters tend to be withdrawn and socially isolated, and consequently experience feelings of loneliness (Vandersall & Weiner, 1970; Heath et al., 1983). It is expected that these youngsters may have a limited number of friends and spend a great deal of time alone. This is consistent with case reports indicating that the majority of firesetting is done alone rather than with one or more acquaintances or friends (Gaynor, 1985). Therefore, it is important to assess the social network of youngsters presenting with pathological firesetting. In addition, they may be questioned about experiencing feelings of isolation and loneliness. They may be unhappy about their social isolation, yet they may lack the appropriate social skills to initiate and sustain meaningful friendships. Teaching youngsters to build and maintain effective social networks may be a target for therapeutic intervention.

School performance and behavior for firesetting youngsters is expected to be below average. Although intelligence may be in the normal range, the presence of learning disabilities is likely to interfere with adequate school achievement. In addition, because the learning disabilities of these youngsters often go undiagnosed, their frustration within learning situations at school often leads them into behavior problems. Many of these youngsters cannot adjust to school environments and are suspended or expelled from several different schools. Hence, it is important to ask parents about school achievement and adjustment to ascertain whether the current school situation is not only appropriate but supportive of the special needs of these youngsters.

The third of the critical elements to be assessed during the clinical interview are those environmental conditions surrounding the reported firestarting. It is likely that there are specific conditions existing within the environment of these youngsters which not only precipitate firestarting, but reinforce the likelihood of its reoccurrence. The three dimensions of environmental conditions are antecedent stressors, behavioral expression, and consequences. For each reported firestart, the specific events occurring prior to the firestart must be analyzed, the unrealized feeling or behavior that may be expressed through the firesetting must be explored, and the consequences occurring as the result of the firestarting must be specified. There may be recurring environmental conditions that set the stage for firesetting behavior. For example, if parents are having marital difficulties and father leaves the house for several days at a time, Frankie may feel resentful and angry and set small fires to draw attention to his feelings. Firesetting incidents often can be linked to significant events occurring within the environment and can be viewed as an expression of unrealized emotion. In addition, if firestarts go undetected or youngsters do not experience appropriate consequences for their firesetting, such as discipline by parents, then their involvement in such activity is likely to continue. Clinicians must evaluate the environmental conditions which family members suggest as significant and which may be related to the pathological firesetting behavior.

The three critical elements of individual characteristics, social circumstances, and environmental conditions and their related dimensions form the core areas to be assessed during the clinical interview. Although individual methods of evaluating these elements may vary, an accurate description of them will reflect the psychopathology underlying firesetting behavior. However, the psychosocial features of youngsters and their families also must be evaluated in light of the severity of the firesetting activity. Hence, the following section will describe the relevant data to be assessed during the clinical interview regarding the history of fire behavior.

Firesetting Behavior History

Crucial to the collection of data during the clinical interview of firesetting youngsters and their families is the ability to evaluate the history of firesetting activity. It is important to understand the developmental process whereby fire interest evolves into patterns of pathological firesetting. Therefore, the characteristics of initial fire interest, the circumstances of fireplay, and details of the reported firestarts must be evaluated to determine the magnitude of the presenting firesetting behavior.

Evidence of fire interest surfaces in the majority of children around the age of three or four. It may take the form of questions about fire-related activities, such as what makes the candles on the birthday cake burn, or it may emerge in play, such as a request for a toy fire truck or toy oven. It is important to discover the age at which youngsters demonstrated their initial interest in fire, how it was expressed, and how it was responded to by parents or other responsible adults. For example, if Lionel asked to touch the fire on the stove to see how hot it was, did mother let him, did mother warn him never to play with the stove, or did mother begin teaching the conditions under which it was safe to experiment and learn about fire? Gauging family reaction to initial fire interest will provide information both on attitudes toward fire as well as efforts made within the home to teach fire safety and fire prevention. Many young children between the ages of six and nine involved in firesetting behavior have not been taught basic fire safety rules. Ruling out lack of fire safety knowledge is essential to determining the etiology of the observed firesetting.

The majority of firesetting youngsters begin their pathological behavior through their involvement in fireplay. However, not all youngsters participating in fireplay become involved in pathological firesetting. Hence, an analysis must be made of the elements of fireplay behavior, where fireplay is defined as those initial firestarts which take place in an unsupervised setting and which are primarily motivated by curiosity and experimentation. The six elements of fireplay that must be evaluated are the age at onset, the method or planning of the firestart, the type of material used in the ignition, what, if anything, was set on fire, the behavior immediately following the firestart, and the consequences of the fireplay incident. The fireplay activity of youngsters occurring at the ages of four or five usually are of an unplanned nature where book matches are typically utilized to ignite paper, trash, or leaves found in or near the home. This scenario may describe the fireplay activities of many youngsters, the majority of whom do not become involved in pathological firesetting behavior. There are factors, however, which distinguish the fireplay activities of normal youngsters from the fireplay activities of those youngsters who are likely to become involved in pathological firesetting.

The elements that characterize the fireplay activity of youngsters likely to become involved in pathological firesetting include the specific behaviors that usually follow their firestarts and the consequences typically experienced after their initial fireplay incident. Once a fireplay incident results in an actual fire, pathological firesetters often will leave the fire scene without trying to extinguish the fire or without notifying someone to help put out the fire. In addition, they frequently return to the fire scene either to watch the fire burn or to see the resulting destruction. Usually the fireplay incident goes unnoticed by adults within the environment, and, therefore, no consequences resulting from fireplay, such as parental discipline or retribution, are instituted. Hence, once pathological firesetters become involved in fireplay, and their fireplay results in an actual fire, they are likely to deny their involvement by failing to respond in socially appropriate or acceptable ways. They will not attempt to extinguish the fire, they will not acknowledge responsibility, and therefore they will not experience the consequences of their firesetting behavior. These elements of fireplay must be assessed during the clinical interview of both the youngsters and their parents. The characteristics of early fireplay behavior are one of the primary indications of what can be expected to occur in future patterns of firesetting.

Firesetting behavior must be evaluated using the six elements of age at onset, the number of firestarts, the method and type of ignition, the intention and target of the fire, the behaviors occurring immediately after the firestarts, and the consequences of the firesetting behavior. It is critical to establish the age at which firestarting began and the number of subsequent firestarting incidents. The age at which fireplay, characterized by a single-episode firestart primarily motivated by experimentation and curiosity, turned into intentional firestarting and the numbers of subsequent firestarts will in part determine the prognosis for correcting the firesetting behavior. Clinical experience suggests that if fireplay moved to intentional firesetting with one or two firestarts occurring before age eight, the prognosis will be better for these youngsters compared to those youngsters over the age of eight who have participated in several planned firestarts. The age of onset coupled with the number of intentional firestarts indicates the strength of the firesetting pattern. The later the age of onset and the greater the number of intentional firestarts, the stronger the firesetting behavior pattern and the more difficult it will be to adjust or change its direction.

Once a determination is made as to the strength of the pattern of firesetting behavior, then each firestart must be individually evaluated. First, the planning of the firestart must be assessed. For example, was the firestart planned over several days with an effort made to pick the best time and place, or was the firestart intentional, but the specifics of time and place not carefully worked out? Second, the type of materials used to start the fires must be specified. They can range in severity from the use of book matches to the use of explosive devices. Third, the stated reason for the fires must be ascertained. This is one of the more difficult pieces of data to evaluate. Many youngsters are unable to verbalize their reasons or offer such answers as "I did it for the fun of it." Given that clinical experience suggests that firestarts frequently are motivated by emotional expression, such as anger or revenge, it might be useful to probe in this direction by offering this as a possible explanation to those who are unable to explain their behavior. Fourth, the object of the fire can provide useful information. If the targets of fires are personal items, this may be an indication of inner struggles or attempts at self-destruction. If the targets are another person's property, this may be an expression of anger or revenge. The meaning of what is set on fire should be recognized and interpreted. Fifth, the behaviors immediately following the firestarts and the experienced consequences also must be assessed. The severity of firestarts increases when there is no attempt to extinguish the fires. In addition, if youngsters are disciplined or experience consequences in the form of retribution and their firesetting continues, then the prognosis for correcting the firestarting behavior and the accompanying psychopathology is likely to be less favorable.

A comprehensive firesetting behavior history will yield specific data concerning the severity of the pattern of firestarting. An analysis of the development of firesetting will indicate the prognosis for adjusting or changing the behavior and the related psychopathology. The prognosis is likely to be more favorable if those circumstances indicated in Table 5.1 characterize the expression of fire interest and the subsequent occurrence of fireplay activity and firestarting behavior. First, if initial fire interest is not met with instruction regarding fire safety, then the communication of fire safety and prevention rules will make a significant difference in fireplay and firestarting motivations. Second, if fireplay activity is characterized by an attempt to extinguish a resulting firestart and appropriate personal and social consequences are experienced as a result of fireplay, then the probability is low that fireplay or firestarting will recur. If fireplay and one or two intentional firestarts occur before the age of eight, the chances are good that the pathological firesetting behavior can be changed. In addition, if the firestarts are ignited by book matches with available materials such as paper, trash, or leaves, and there is an attempt to extinguish the fire, accompanied by feelings of remorse and guilt, then the prognosis is excellent with respect to preventing the occurrence of future

Table 5.1
Circumstances of Fire Behavior Indicating a Favorable Prognosis

Behavior	Prognostic Indicator
Fire interest	• Must be met with education and instruction in fire safety and prevention.
Fireplay	• Attempts are made to extinguish unintentional firestarts. • Appropriate retribution and consequences are experienced.
Firesetting	• Early onset, before age eight. • No more than two intentional firestarts within a six-month period. • Use of available materials, such as book matches, for ignition. • Targets of firestarts are available materials such as paper, trash, or leaves. • Attempts are made to extinguish firestarts. • Feelings of guilt or remorse are experienced subsequent to firestarts.

firestarts by implementing the appropriate intervention strategy. Although intentional firestarting is a seriously dangerous antisocial activity, if the specific elements characterizing the circumstances of firesetting are evaluated, then the probability of recurrence of the pathological behavior can be estimated.

A complete clinical interview plan must contain an analysis of the dimensions comprising two major classes of data—the three critical elements identifying presenting psychopathology, including individual characteristics, social circumstances, and environmental conditions, and the history and pattern of firesetting behavior. Although there are various methods clinicians may choose in collecting and analyzing these data, the result should be an accurate clinical description of existing psychopathology coupled with a detailed enumeration of the severity of the firesetting behavior pattern. These two major classes of data will be synthesized to identify targets of behavior change, establish a working clinical diagnosis, and design an effective strategy of intervention.

BEHAVIOR, DIAGNOSIS, AND INTERVENTION

There are three primary outcomes of the clinical interview of firesetting youngsters and their families. First, the clinical data collected on the critical elements identifying psychopathology and the history and pattern of firesetting behavior can be synthesized to target those behaviors which must be adjusted or changed. Second, the data can be utilized to establish a working clinical diagnosis. Third, based on the targeted behaviors to be changed and the clinical diagnosis, an intervention strategy can be designed to eliminate firestarting and adjust those factors responsible for the underlying psychopathology accompanying the firesetting behavior.

Targets of Behavior Change

An analysis of the clinical data pertaining to the three critical elements of individual characteristics, social circumstances, and environmental conditions and their related dimensions will identify behaviors that contribute to the psychopathology associated with firesetting. These behaviors must be recognized and evaluated during the clinical interview. These are specific categories of behavior which must be defined as accounting for the observed psychopathology that accompanies the firesetting behavior. These behaviors must become the targeted areas for adjustment and change to eliminate firestarting and remediate the associated psychopathology.

The predominant behavior targeted for change is firestarting. Surprisingly, the specific behavior of firesetting is relatively easy to eliminate. Future chapters will describe effective methods to stop the occurrence of this behavior. Unfortunately, firestarting does not usually occur as an isolated behavior. These are psychopathological characteristics contributing to firesetting behavior which are resistant to adjustment or change. Hence, through the clinical interview it is essential that specific targets of behavior change be identified and accurately described within the context of individual characteristics, social circumstances, and environmental conditions.

There are three targets of behavior change—physical, cognitive, and emotional—which must be considered within the critical elements of individual characteristics. Energy level, with indications of hyperactivity, evidence of enuresis, and history of abnormal sexual activity, including abuse, comprise potential areas of change for physical behavior. Cognitive functioning includes the presence of learning disabilities as possible targets of behavior change. The experience and expression of emotions such as anger, frustration, and revenge are targets of potential adjustment. The clinical interview should produce an accurate assessment of these behaviors and the identification of specific characteristics targeted for change.

Family, peer relationships, and school activity are the three targets of behavior change comprising social circumstances. Adjustments can be made with respect to family structure and patterns of interaction, which include role modeling and methods of guidance, supervision, and discipline. If parental psychopathology is evident, intervention recommendations also must be considered. In addition, the number and quality of interpersonal relationships can be targets for behavior change. School performance and behavior can also be adjusted. In general, social circumstances may be less resistant to change than individual characteristics and, once adjusted, may contribute to a significant reshaping of the psychopathology accompanying firesetting behavior.

The environmental conditions targeted for change include behavioral expression and consequences. An assessment of the emotions and behaviors expressed through firesetting and how they may be experienced and expressed in less destructive and more socially acceptable ways marks an important area of adjustment. In addition, if future firestarting incidents occur, a clear understanding must be specified of the impending consequences, including parental discipline, and legal, financial, and intervention options that may be implemented. The ability to adjust both social circumstances and environmental conditions may contribute most to a significant change in the psychopathology that sustains patterns of firesetting behavior.

Identification of targets of behavior change represents a synthesis of the data collected on critical elements and firesetting behavior during the clinical interview. An analysis of the behaviors to be adjusted or changed should lead to formulation of a working clinical diagnosis. In addition, once targets of behavior change have been identified and a working clinical diagnosis has been established, then an effective strategy for intervention can be recommended.

Clinical Diagnosis

The data collected during the clinical interview on the critical elements of psychopathology and the subsequent identification of targets of behavior change can be utilized to establish a working clinical diagnosis, first by ruling out the possibility of an existing psychosis, and second by selecting an appropriate diagnosis from the emotional or behavioral classes of psychiatric disturbances. During the clinical interview with parents and their youngsters, clinicians must keep in mind that they will be formulating a working diagnosis from their clinical impressions. Although the diagnosis will be ascertained for youngsters presenting with pathological firesetting, it is important also to assess the potential for psychiatric disturbances among parents and other family members participating in the clinical interview.

Clinical evidence indicates that the probability of firesetting youngsters being diagnosed as psychotic is low (Gaynor, 1985). Nevertheless, the essential features characterizing psychosis must be actively ruled out by an analysis of the presenting clinical data on the critical elements of individual characteristics and social circumstances and the motivation underlying the firesetting behavior.

Childhood psychosis most frequently is characterized by the presence of one or more of such features as disturbed thought content, altered perceptions, inappropriate or flat affect, and extreme withdrawal. A clinical assessment of cognitive functioning must rule out the presence of disturbed thought content and altered perceptions. An evaluation of emotion should reveal the absence of a severe disorder, although it is not unusual to find firesetting youngsters emotionally unexpressive and feeling little guilt regarding their firestarting behavior. In addition, the assessment of the stated reasons for firesetting should not include such explanations as "the voices in my head made me do it" or "my god told me to burn down my earth and I would be a hero." An analysis of firesetting motivation should indicate the absence of bizarre thought content regarding reasons for firesetting. If the clinical data on the individual characteristics of cognitive and emotional functioning are not indicative of psychosis, and the motivations for firesetting reveal no bizarre thought content, then a clinical diagnosis of psychosis can be ruled out. Hence, the remaining psychiatric disturbances of emotional and behavioral disorders must be considered potential clinical diagnoses.

The two most frequently diagnosed emotional disturbances for firesetting youngsters are schizoid and oppositional disorders. If an analysis of the targets of behavior change identifies the unique and most significant area for adjustment as the ability to initiate and sustain interpersonal relationships, then the diagnosis of schizoid should be given serious consideration. The occasional outbursts of aggressive behavior that characterize this disorder are manifested in the exhibited firesetting behavior. If there is evidence of emotional problems described by negativistic, provocative behavior toward authority, accompanied by expressions of anger, and an analysis of the targets of behavior change indicate poor social adjustment and school performance, then the diagnosis of oppositional disorder is likely to be appropriate.

If the data collected during the clinical interview rule out the possibility of psychosis and the diagnoses of schizoid and oppositional disorders are not evident, then the behavioral disturbances of conduct and attention deficit disorders are likely to be the preferred clinical diagnoses. Clinical and empirical studies demonstrate that firesetting youngsters are most likely to be diagnosed with these disorders (Siegelman & Folkman, 1971; Kuhnley et al., 1982; Stewart & Culver, 1982; Heath et al., 1983; Kolko et al., 1985), with the primary diagnosis being conduct disorder, which may be accompanied by the secondary diagnosis of attention deficit disorder. The essential feature of a conduct disorder is a persistent pattern of aggressive behavior violating the basic rights of others. This feature describes the presenting problem of pathological firesetting. In addition to interpreting firesetting behavior as a violation of the safety of others, the aggressive component of the diagnosis, characterized by physical violence against property, fits the behavior pattern of firesetting. If an analysis of the targets of behavior change reveals deficits concentrated within the critical element of social circumstances, indicating significant family problems focused on inappropriate use of harsh punishment and evidence of parental psychopathology, coupled with poor school performance and behavior, then these features confirm the conduct disorder diagnosis. If there is a failure to establish a normal degree of affection, empathy, or significant relationships with others, then the subtype of undersocialized also must be considered. The characteristics describing the majority of recurrent firesetting youngsters most often fit the essential features of the diagnosis of conduct disorder.

If, in addition to the presence of the essential features characterizing the diagnosis of conduct disorder, an analysis of the targets of behavior change indicates deficits within the physical and cognitive dimensions of individual characteristics, then the possibility of a secondary diagnosis must be entertained. If there is evidence within the cognitive dimension of a shortened attention span accompanied by an indication within the physical dimension of a high energy level, then an attention deficit disorder diagnosis may be warranted. In addition, if there is evidence of poorly organized gross motor activity, then the attention deficit disorder may include a diagnosis of hyperactivity. The behavior of firesetting youngsters generally presents as a conduct disorder, with the secondary diagnoses of attention deficit disorder and hyperactivity occurring with less regularity.

A working clinical diagnosis is arrived at by summarizing the data collected during the clinical interview, identifying the targets of behavior change and synthesizing this information to coincide with those essential features characterizing psychiatric disturbance. Once the possibility of psychosis is ruled out, then a primary clinical diagnosis can be selected for youngsters presenting with firesetting behavior from the two classes of emotional and behavioral disorders. The most frequently occurring primary clinical diagnosis is conduct disorder. Often there may be accompanying symptoms suggesting a secondary diagnosis. An analysis of the targets of behavior change and establishing a clinical diagnosis leads directly to the design of an effective intervention strategy.

Intervention

Recommendations for intervention must emerge from identification of the targets of behavior change and the implications of the clinical diagnosis. The primary target for change is the firesetting behavior. Youngsters involved in firesetting display significant psychopathology which also needs immediate attention. The question becomes how to stop the firesetting behavior and adjust or change the accompanying psychopathology. There are two primary considerations. First, the behaviors in which the psychopathology is manifested must be identified and targeted for change—hence, the development of targets of behavior change. Second, methods must be proposed to facilitate change in the targeted behaviors. Therefore, the final step in the evaluation of firesetting youngsters is to utilize the clinical data to recommend an intervention strategy to stop the firesetting and adjust or change the underlying psychopathology which sustains the behavior.

The organization of the data collected during the clinical interview by the three critical elements of individual characteristics, social circumstances, and environmental conditions and their associated dimensions provide the framework for identifying the psychopathology that accompanies firesetting behavior. When the clinical data are analyzed according to this framework, the result is identification of the specific targets of behaviors related to the observed psychopathology of firesetting. These are the behaviors targeted for change. If they can be adjusted successfully, then the probability is low that firesetting will reoccur. If these targeted behaviors cannot be adjusted or changed, then the likelihood is high that firesetting or other antisocial behaviors will resurface. Hence, identification of targets of behavior change lays the foundation for recommending specific methods of intervention.

Although each case of firesetting will have its own unique set of behavioral characteristics, aside from the common objective of eliminating the specific behavior of firestarting, there are a set of targets of behavior change which describe a significant number of recurrent firesetting youngsters. Table 5.2 enumerates the most frequently occurring targets of behavior change characterizing youngsters presenting with pathological firesetting. The physical, cognitive, and emotional dimensions of individual characteristics are most frequently targets for change in firesetting youngsters. High levels of energy, enuresis, and sexual abuse are behaviors that most often need to be adjusted within the physical dimension. The existence of learning disabilities, especially in the areas of concentration and ability to pay attention to direction, typifies cognitive deficits. The overwhelming experience and subsequent inappropriate expression of anger often is the focus of emotional readjustment. Modifying patterns of family interaction, especially with respect to the types of role modeling and guidance, supervision, and discipline methods, improving the number and quality of interpersonal relationships, and correcting school performance and behavior are frequently targets for change within social circumstances. Finding avenues of appropriate emotional and behavioral expression other than firesetting and implementing appropriate consequences when firestarting behavior occurs are the two environmental conditions that

Table 5.2
Targets of Behavior Change for Firesetting Youngsters

Target	Dimension	Behavior
I. Individual characteristics	A. Physical	1. Energy level
		a. Hyperactivity
		2. Enuresis
		3. Sexual abuse
	B. Cognitive	1. Learning disabilities
	C. Emotion	1. Overwhelming experience
		a. Anger
		b. Aggression
		2. Inappropriate expression
		a. Anger
		b. Aggression
II. Social circumstances	A. Family	1. Structure
		2. Behavior
		a. Environment
		b. Role modeling
		c. Guidance, supervision, and discipline
		d. Physical abuse
		3. Parental psychopathology
	B. Peers	1. Social skills
	C. School	1. Performance
		2. Behavior
III. Environmental conditions	A. Behavioral expression	1. Anger
		2. Attention
	B. Consequences	1. Implementation

must be adjusted for firesetting youngsters. Clinical experience suggests that the set of behaviors must be the targets of change for the design of an effective strategy of intervention.

Once the targets of behavior change have been identified for firesetting youngsters, then methods for implementing the expected adjustments or changes must be designated. A strategy for the immediate elimination of firestarting behavior must be recommended. Specific therapeutic methods also must be considered to remediate the psychopathology evident for each of the behaviors targeted for change. Interventions must be suggested to adjust the observed deficits in physical, cognitive, and emotional functioning. Methods for changing patterns of family interaction and improving interpersonal relationships must be indicated. If school performance and behavior need improvement, then the appropriate steps for change must be identified. Those environmental conditions sustaining the patterns of firesetting must be modified. Methods of intervention must be proposed which adjust or change the targeted behaviors associated with the psychopathology of firesetting.

The specific types of interventions designed to eliminate firestarting and remediate the accompanying psychopathology are discussed in the next chapter. The recommended intervention strategy for helping firesetting youngsters will be effective primarily if it emerges from a thorough clinical interview which evaluates the critical elements associated with the presenting psychopathology and the history and pattern of firesetting behavior, analyzes them in terms of targeted behaviors which must be changed, and establishes a working clinical diagnosis. If these procedures are followed, then firesetting youngsters and their families will be given the best chance to help themselves correct and prevent the continuation of a seriously disturbing and dangerous behavior pattern.

SUMMARY

The clinical interview is offered as the best method of conducting an evaluation of firesetting youngsters and their families. Some observations are made regarding circumstances specific to conducting an assessment where firesetting is the presenting problem. Options are suggested for the interview format, including who will be interviewed and how, the length of the interview, and the expected product of the evaluation. The preconceived attitudes that youngsters, parents, and clinicians bring to the interview, such as their hopes, fears, and fantasies, are described for the purpose of adjusting expectations to a realistic level. Various styles of interviewing are recommended to maximize the collection of clinical data while minimizing the amount of anticipated or realized stress associated with the evaluation. It is suggested that the clinical evaluation follow an interview plan. This interview plan is comprised of two major classes of data—those critical elements such as individual characteristics, social circumstances, and environmental conditions, which identify psychopathology and those critical elements such as the characteristics of fire interest, fireplay, and firestarting which describe the history and pattern of firesetting behavior. Specific methods are recommended for collecting these data during the clinical interview. The analysis and synthesis of these clinical data are shown to identify targeted behaviors indicative of existing psychopathology that must be changed, establish a working clinical diagnosis, and design an effective intervention strategy to help firesetting youngsters and their families change the course of a serious and dangerous pattern of behavior.

6

Pathological Firesetting and Psychotherapy

The goals of an effective intervention strategy are to eliminate pathological firesetting behavior and sustain significant changes in those target behaviors reflecting the accompanying psychopathology. During the last few years, psychotherapies have been developed to work specifically with youngsters and their families presenting with firesetting as a primary behavior problem. Outpatient and inpatient intervention are the two treatment approaches that stop pathological firesetting and adjust the accompanying psychopathology. Cognitive-emotion, behavior, and family psychotherapy are the three predominant methods of outpatient therapy. Psychodynamic and behavior therapy represent the two major methods of inpatient intervention. All these methods will be described in terms of their underlying psychological philosophy or theory, the treatment modalities employed, the types of direct interventions utilized to eliminate firesetting, and the techniques applied to change specific targeted behaviors representing associated psychopathology. Because the majority of these intervention strategies are recently developed and implemented, there is an absence of controlled clinical studies demonstrating their relative effectiveness. However, there have been some clinical attempts to assess the therapeutic effectiveness of these strategies in eliminating firesetting behavior and adjusting the accompanying psychopathology, and these evaluations will be presented as a preliminary method for determining the value of these interventions. An effective treatment outcome primarily depends on the optimal match between the accurate identification of the behaviors targeted for change and the selection and application of the appropriate intervention strategy.

OUTPATIENT TREATMENT

The salient outpatient treatment of choice for firesetting youngsters is psychotherapy. There are two primary components to be considered in developing an outpatient psychotherapy program for firesetting youngsters. The first is the modality or who will participate in the psychotherapy. The second is the method or the technique employed within the therapeutic setting. A variety of modalities and methods of psychotherapy have been applied to the outpatient treatment of firesetting youngsters.

The two predominant modalities utilized in the outpatient treatment of firesetting youngsters are individual and family psychotherapy. The primary focus of individual psychotherapy is the immediate cessation of firesetting, with a secondary emphasis on adjusting or changing the accompanying psychopathology. The methods of cognitive-emotion and behavior therapy most frequently employ the modality of individual treatment. The participants in individual psychotherapy usually are the youngsters exhibiting firesetting behavior; however, there have been a growing number of case reports describing individual behavior therapy techniques targeted at teaching parents how to eliminate firesetting in their youngsters. Although the application of family psychotherapy to the problem of firesetting is reported less frequently, the clinical literature suggests that this modality addresses both the cessation of firesetting as well as the underlying psychopathology. The methods of family psychotherapy are dynamic in orientation and stress improving patterns of communication and interaction. The modalities of individual and family outpatient treatment will be described within the context of the specific methods of psychotherapy utilized to treat firesetting youngsters.

The three methods of outpatient treatment that will be presented and evaluated are cognitive-emotion therapy, behavior therapy, and dynamic family psychotherapy. An innovative clinical report will be outlined describing the application of cognitive-emotion therapy to treat firesetting youngsters in individual psychotherapy. The focus of this therapeutic method is cessation of firesetting through the association of cognitive and emotional experiences with actual firesetting activity. Clinical reports will be reviewed detailing specific behavior therapy techniques designed to eliminate patterns of firesetting. Behavior therapy methods will be described that utilize both youngsters and parents as instruments in stopping firesetting behavior. The dynamic family psychotherapy method of treating the problem of firesetting will be reviewed from case studies suggesting that adjusting the underlying maladaptive patterns of social interaction within the family will remediate the overt symptom of firesetting. Finally, each of these three methods of outpatient treatment—cognitive-emotion, behavior, and family therapy—will be evaluated with respect to its effectiveness in eliminating firesetting behavior and the accompanying psychopathology.

Cognitive-Emotion Psychotherapy

The application of cognitive-emotion psychotherapy to the outpatient treatment of firesetting youngsters has recently been reported as successful in eliminating firesetting in 27 of 29 youngsters (Bumpass, Fagelman, & Brix, 1983). The major goal of this psychotherapy is to bring firesetting behavior under the control of youngsters. This is accomplished by utilizing a "graphing" technique to bring to awareness the events and feelings associated with firesetting activity. There is a typical or expected pattern of feelings which therapists try to elicit from youngsters. Once these events and feelings are realized, the majority of youngsters are able to recognize the emergence of their urge to firestart, interrupt the behavior before it starts, and substitute socially appropriate types of behaviors to express their underlying emotions.

The primary mechanism of this psychotherapy is construction of a written graph by youngsters with guidance from the therapist. Often parents are invited to participate in the process. Both the feelings and specific events leading up to and following the most recent firestart are graphed. The horizontal axis of the graph represents the time period before, during, and after firesetting, and the vertical axis represents the intensity of the emotions experienced during this time period. Youngsters are first asked to list the significant events, such as school holiday, mother left house to run an errand, looked for matches, lit fire, and ran away and hid, which occurred surrounding the firestarting incident. These events are recorded in their sequence of occurrence along the horizontal axis of the graph, with the firestarting episode placed in the middle of these events. The youngsters next are asked to describe their patterns of feelings associated with these events. Each type of feeling such as happiness, loneliness, or anger is graphed separately during this time period and is represented by an individual line. Because specific feelings will wax and wane during this time period, their various intensity levels will be represented by the relative amplitude of the "feeling" lines on the graph. For example, if Johnny initially felt happy about having a school holiday, but felt less so when his mother left him alone in the house while she ran an errand, then the amplitude of the happiness feeling line will first be high, representing a greater intensity of happy, and then will be relatively lower, indicating a less happy feeling state. In this way, one or more feelings and their relative patterns of experienced intensity are graphically represented in relationship to the significant firestarting incident.

A typical or usual graph is expected for the majority of youngsters involved in firesetting. Although the specific events surrounding the firestarting may vary, the specific pattern of feelings is expected to be similar for these youngsters. It is the role of the therapist, through the process of constructing this graph with youngsters, to bring to their awareness the usual feeling states associated with the impulse of firestarting. The typical graph indicates that one or more significant events trigger a sequence of sad, lonely feelings. These feelings are replaced by intense, angry feelings, which are in turn controlled by a destructive urge that is significantly relieved by setting a fire. A feeling of fear usually emerges before the firestart and continues for a brief period of time after the fire is set. Feelings of guilt, if experienced, will follow the firestart. The therapist emphasizes the importance of the relationship between these experienced feelings and the desire to set a fire. This pattern of feelings, the associated significant events, and firestarting are the focus of the psychotherapy.

Through the remainder of psychotherapy, the therapist emphasizes that youngsters can prevent themselves from setting more fires. First, they are told that the feelings they experience early in the typical pattern leading to firesetting, such as sadness and loneliness, are a signal that the impulse to firestart may be imminent. It is suggested to these youngsters that they do not always act on their feelings, especially if they are likely to result in destructive activity. Youngsters also are told that they have a choice of what they do about their feelings. They are asked what constructive behaviors they might want to pursue when they begin to feel sad or lonely. These alternative behaviors are listed on the graph. Youngsters are told that they will probably not set any more fires, but if they have the urge to firestart, they should pay attention to their feelings so that they can talk about them with the therapist. If youngsters become overwhelmed by their feelings and want to set a fire, they are encouraged to first telephone the therapist. If subsequent firestarts occur, then these firesetting incidents are graphed and become the focus of future sessions. If youngsters are successful in redirecting their destructive urges, they are supported and praised. The therapist emphasizes that sad, lonely, and angry feelings are experienced by nearly everyone; however, it is the behaviors that accompany these feelings which must be expressed in a socially acceptable manner.

This psychotherapeutic approach stops firesetting by assuming that the behavior is the result of the occurrence of environmental conditions, namely, the antecedent stressors of significant events that trigger a specific pattern of feelings, which, in turn, sets the stage for firestarting. Firesetting occurs when feelings reach an unacceptable intensity and some type of behavioral expression is necessary. By graphing this sequence of feelings and events, youngsters are taught how to recognize the early onset of this pattern, interrupt it, and replace a potential destructive feeling and accompanying behavior with a socially acceptable response. Hence, not only does the firestarting stop, but additional targeted behaviors representing associated psychopathology are adjusted. This therapeutic method targets individual characteristics and environmental conditions as important areas of behavior change. Within individual characteristics, the dimension of emotion is acknowledged as important, and the tendency to be overwhelmed by feelings as well as to become involved in the inappropriate expression of these feelings is the major focus of this therapeutic intervention. Within environmental conditions, the dimension of antecedent stressors, represented by the occurrence of significant events, and the dimension of behavioral expression, with firestarting interpreted as the result of intolerable feelings such as anger, are both salient features within the therapeutic process. This cognitive-emotion approach to the outpatient treatment of firesetting youngsters stops firestarting and works on aspects of individual characteristics and environmental conditions as targets of behavior change.

For all 29 youngsters reported to have participated in cognitive-emotion psychotherapy, there was an attempt to conduct a follow-up assessment on the effectiveness of the treatment in eliminating their firesetting behavior and the accompanying psychopathology. Twenty-six of the twenty-nine youngsters were assessed by telephone interview at various follow-up time intervals ranging from six months to eight years, with an average duration of two and one-half years. The data from the follow-up assessment indicated that 2 of the 29 youngsters exhibited single episodes of firesetting behavior subsequent to the termination of treatment. For six youngsters, additional psychotherapy was recommended once the graphing technique had been successful in stopping their firestarting behavior. However, treatment was not actively pursued, and consequently these youngsters became involved in other acting-out and antisocial behaviors such as theft and vandalism. They were subsequently placed in either outpatient or residential treatment programs. Although there was no control group included in this evaluation, the authors argue that the youngsters served as their own controls in that all of these families had attempted to overtly control the firesetting behavior, and many had sought treatment unsuccessfully for firesetting prior to their participation in cognitive-emotion psychotherapy.

Initial follow-up evidence suggests that cognitive-emotion therapy utilizing the graphing technique shows promise in eliminating firesetting behavior and the accompanying psychopathology. The application of this outpatient intervention method and its relative effectiveness deserve further attention. In the future, questions must be asked regarding both the type of youngsters and the severity of the firesetting behavior most successfully treated utilizing this type of psychotherapy. For example, it may be that the graphing technique works most effectively with younger children between the ages of six and nine who have been involved in several intentionally set fires which caused significant damage to property. In addition to understanding the types of youngsters and the severity of behavior most successfully remediated by this psychotherapy, it would be useful to compare its relative effectiveness to other outpatient treatment methods such as behavior therapy. The answers to these questions will clarify the long-term impact of cognitive-emotion psychotherapy in the successful treatment of pathological firesetting.

Behavior Therapy

Several case studies have been reported in the literature applying behavior therapy techniques to the outpatient treatment of youngsters presenting with the problem of firesetting. The predominant behavior therapy methods employed either alone or in combination are punishment, reinforcement, negative practice or satiation, and operantly structured fantasies. Four case studies will be presented describing the application of these behavior therapy techniques to the outpatient treatment of firesetting. These case reports were chosen because they are representative of the entire range of behavior therapy methods employed in the treatment of pathological firesetting. In addition, all of these case studies report follow-up data so that questions can be addressed regarding the effectiveness of these behavior therapy interventions in eliminating firesetting behavior and the accompanying psychopathology.

Carstens (1982) reports that the firestarting behavior exhibited by a four-year-old boy stopped when a work penalty system utilized as a threat of punishment was set into place by the parents. This youngster had a short, intense history of unsupervised match play, with efforts made by his parents to encourage the striking of matches in their presence. In addition to match play, this youngster attempted to set fire to his own and his parents' bed. Both attempts were aborted when his activity was discovered by his parents.

The work penalty system was presented to this youngster and his parents by the therapist. It consisted of one hour of hard labor if the parents found misplaced matches or lighters in the home or discovered their son in possession of them. He was to be presumed guilty of match play under these circumstances even if he did not actually strike the match. The work penalty system involved one hour of closely supervised hard labor and included such activities as scrubbing off the back porch, washing the walls, or cleaning the space between the kitchen tiles with a toothbrush. Once the threat of employing work penalty as a form of punishment for match play and firestarting was established within the family system, there were no further incidents of firesetting behavior.

A six-month follow-up assessment was conducted by the therapist with the parents. They reported that their son no longer was interested or participated in match play or firestarting. Hence, the threat of punishment, which included a detailed description of the nature and administration of the consequences, was a sufficient deterrent to future firesetting behavior. The threat, not the actual implementation of punishment, appeared to be the contingency that eliminated the recurrence of match play and nonproductive firestarting behavior.

The threat of punishment also has been employed in combination with a program of positive reinforcement. Holland (1969) reports the cessation of firestarting in a seven-year-old boy, the oldest of three children, who had engaged in fireplay once or twice a week for at least a three-month period. The parents participated in behavior therapy sessions which taught them how to control the firesetting behavior of their son.

The presenting problem of the parents was viewed by the therapist as not only involving the suppression of their son's undesirable behavior through the effective use of punishment, but also as reinforcing socially acceptable behaviors associated with nonproductive firesetting, such as bringing found matches to parents and controlling the urge to strike matches in an unsupervised setting. The threat-of-punishment procedure utilized by the parents involved warning their youngster that if he engaged in unsupervised firestarting, his new and most prized possession—a baseball glove—would be taken away and physically destroyed in his presence. At the same time they introduced the threat of punishment, the parents employed a program of positive reinforcement.

The program of positive reinforcement consisted of three major steps. The first step was to instruct their son to bring matches found in the home immediately to his father. Meanwhile, the father conspicuously hid one empty matchbook. When the youngster found it, he returned it to his father, who immediately gave him five cents to spend any way he wanted. Next, the father hid additional matches and matchbooks, the youngster found them and returned them for monetary rewards. The youngster was put on a continuous reinforcement schedule for eight trials, varying the degree of reinforcers, from 1 to 10 cents. For the next few trials, the youngster was told he was not to expect money every time as a reward. This procedure was determined successful when retrieval was occurring at such a high rate that the youngster was saving matches and matchbooks found during the day to give to his father.

Since there always exists the possibility that the youngster would find matches outside the home when parental reinforcement was not available, the second step, strengthening nonstrike behavior, was introduced. The youngster was asked to strike books of 20 matches in front of his father. For every match he did not strike he received a penny. For each book of matches, 20 promised pennies were placed in front of the youngster. If a match was struck, one penny would be removed. Each trial consisted of one book of matches. By the third trial the youngster did not strike any matches.

For the next few trials he was told that he was not going to know what he would receive if he did not strike a match. The reward schedule was varied from no cents to 10 cents. The youngster was successful in not striking matches regardless of the reward schedule.

The final step of the positive reinforcement program was to give social reinforcement, such as praise, along with, and eventually instead of, monetary rewards. Over the next eight months, the parents reported that their son did not exhibit any firestarting behavior at home or in the neighborhood. The positive reinforcement program prevented the implementation of the plan for punishment. At one-year follow-up there were no further reports of nonproductive firestarts.

The structured or planned striking of a match for the purpose of reducing the reinforcing properties and frequency of firestarting, referred to as negative practice or satiation, has been reported successful in eliminating firesetting in several clinical cases without utilizing the additional threat of punishment (Welsh, 1971; Jones, 1981). Negative practice accompanied by corrective consequences and the application of a token reinforcement program has been shown to be effective not only in stopping pathological firestarting but in replacing aggressive, antisocial behaviors with more compliant, socially appropriate behaviors (Kolko, 1983). These methods were successful in treating a six-year-old, developmentally disabled boy presenting with a history of four significant firestarts, the most serious of which consisted of intentionally placing a puppy inside his grandfather's van and then burning it to the ground. The therapist taught the youngster's mother to implement the behavior therapy methods of negative practice with corrective consequences and token reinforcement to eliminate her son's firesetting and adjust some of his socially aggressive behaviors.

Although in the previous case it was demonstrated that positive reinforcers can strengthen nonstriking behavior in controlled firestarting conditions, the method of negative practice with corrective consequences allows the youngster not only to experience but to be responsible for the naturally occurring unpleasant results of a fire. It is anticipated that being responsible for these consequences will satiate the youngster's need to firestart. To implement this method, the mother instructs the youngster to collect a pack of matches, two sheets of paper, a large metal basin, a pail of water, a water hose, a scrub brush, and dishwashing liquid. After the equipment is properly placed, the mother helps her son ignite the paper in the metal basin and extinguish the small fire with the water hose. The corrective consequences for firesetting are that the youngster must clean the basin using detergent and rinse it with water. During this clean-up time the mother reviews this training procedure and discusses the importance of fire safety rules. This negative practice and clean-up procedure is carried out every day after school for four weeks. For the next two weeks, the procedure is carried out every other day, with the mother asking her youngster on the alternative days whether he felt as if he wanted to light a fire. If he said no, he was praised, but if he said yes, the procedure was carried out. When asked, if the youngster responded that he did not want to light a fire, then the negative practice procedure was reduced to once a week and finally eliminated altogether.

To replace the firesetting behavior with socially appropriate behavior, a token reinforcement system was established by the mother. For each day that a fire was not set, and a socially acceptable behavior was exhibited, such as cooperative play with his sister, the youngster received a piece of a puzzle. When all the pieces of the puzzle were collected and no fires had been set, he was entitled to select a backup reinforcer. Puzzles with increasing numbers of pieces were administered over time. The backup reinforcement system was gradually replaced by social reinforcement. The youngster successfully demonstrated an increasing number of socially acceptable and appropriate behaviors.

Fifteen months after the termination of treatment, the mother was contacted for a follow-up assessment. The mother reported that her youngster had not set any fires since their participation in the treatment program. In addition, there had been a dramatic decrease in the number of fighting incidents between the boy and his sister, and this level had been maintained during the follow-up period. This case supports the effectiveness of negative practice with corrective consequences coupled with a token reinforcement system to stop firesetting and replace previously exhibited antisocial behaviors with those of a more socially appropriate nature.

The application of a positive reinforcement program with the implementation of operantly structured fantasies also has reportedly been successful in eliminating the firesetting behavior of a seven-year-old boy (Stawar, 1976). Living with his mother, this young boy participated in four firesetting incidents within a three-month period. Upon discovering these activities, his mother discussed fire safety rules and punished him for his behavior, but he continued firesetting. The therapist first worked with the youngster employing an operantly structured fantasy procedure to control impulsive firestarting and then included the mother in treatment to help implement a positive reinforcement program to prevent the recurrence of firesetting in the home.

The purpose of utilizing the operantly structured fantasy method is to teach the youngster the correct behavioral response when he finds matches or other sources of ignition. This youngster was seen for two 50-minute sessions during which he learned the following story.

> *The Boy Who Found Some Matches*. Once upon a time, a little boy was playing in the yard (the location was changed on subsequent versions to include all rooms of the house, the school, and other places the child frequently visited). While he was playing, the boy just happened to find some matches. Well, this was a very good little boy, and he knew just what you are supposed to do when you find matches. He took them to his mother right away. (In other versions other adults were used.) His mother was very proud of him for bringing the matches so quickly, without striking any of them. His mother told him that he was a very good boy who could act very grown up. Mother then took the matches and put them away and gave the little boy some of his favorite candy for bringing the matches to her.

The youngster learned the story by play-acting with family dolls and a large, open dollhouse. He was asked to tell the story back after he had heard it twice. He was prompted and asked questions, and when he responded with the correct sequence of the story, he was rewarded with a penny candy. In addition, when he recited those parts of the story in which the little boy responded correctly upon finding the matches, he was reinforced with penny candy. At the beginning of the second 50-minute session, the youngster was able to tell the story with only partial reinforcement, and by the end, he remembered it perfectly.

Once the youngster learned the story, the operantly structured fantasy procedure was considered successful. The next step was to involve the mother. The mother was told that her son now knew what to do when he found matches. She was told that if her son found matches and brought them to her, she was to praise him, put the matches away, and give him a penny candy. The mother also was given a copy of the story. She then was instructed to place an empty book of matches where her son would be sure to find them twice a day for the next two weeks. If he responded correctly, he was to be rewarded, and noncompliance was to be ignored. She was asked to keep a written record of the trials and her son's responses.

During the two-week period, 23 trials took place and the youngster successfully redeemed 19 empty matchbooks. There were no reported attempts of firestarting. One- and seven-month follow-ups confirmed that no further firesetting behavior was evident. This behavior therapy method emphasized that socially appropriate responses, such as what a youngster must do when he finds a book of matches, can be taught by utilizing a system of positive reinforcement to shape both cognitive as well as behavioral responses.

All of these reported case studies demonstrate the success of various types of behavior therapy techniques in eliminating pathological firesetting. Regardless of the combination of methods—the threat of punishment, positive reinforcement programs, negative practice and satiation procedures, or operantly structured fantasies—they all were successful in stopping nonproductive firestarting in single case applications. Several questions remain open regarding the effectiveness of these behavior therapy approaches. First, these techniques must be applied to the treatment of more than one youngster before their success can be fairly evaluated. Second, it is unclear which behavior therapy methods are effective with which type of youngsters. The threat of punishment alone may work well in eliminating the early firestarting behavior of one group of youngsters, whereas it may not deter the longstanding patterns of firesetting in another group of youngsters. Third, because youngsters presenting with firesetting behavior often exhibit additional psychopathology, it is necessary not only to stop the observed firestarting, but to treat the underlying psychosocial determinants and associated problematic behaviors. Only two of the four case studies mention pathological behaviors accompanying firesetting, and only one study (Kolko, 1983) describes a positive reinforcement program to replace aggressive, antisocial behaviors with socially appropriate behaviors. Hence, the success of behavior therapy methods in treating the psychopathology that often accompanies firesetting remains to be assessed. Finally, the effectiveness of behavior therapy versus other methods of psychotherapy in correcting patterns of firesetting and the accompanying psychopathology is a topic for future research and evaluation on the relative success of various strategies of outpatient intervention.

Family Psychotherapy

Three significant cases are reported in the literature applying outpatient family psychotherapy to the treatment of pathological firesetting (Eisler, 1974; Minuchin, 1974; Madanes, 1981). In all three cases, the underlying theory is that the emergent problem, namely firesetting, reflects a more generalized disturbance within the family system. Hence, treatment focuses on correcting the maladaptive relationships within the family. Once this is accomplished, the symptom of the family conflict, which in these cases is identified as firesetting, is remediated. Although each of these three cases utilizes different family therapy techniques, they all employ a common set of goals which include an analysis and adjustment of role relationships and communication patterns within the family network.

In two of the three cases the task of teaching family members how to safely start and extinguish fires is utilized as the vehicle to restructure the existing interpersonal relationships and communication channels within the family (Minuchin, 1974; Madanes, 1981). In the first case, a single-parent family, consisting of mother and four children aged 10 and under, seeks treatment when the third-born child, a seven-year-old girl, exhibits multiple episodes of fire-setting. Her most recent incident involved setting fire to the mattress on her mother's bed. Her eight-year-old sister discovered the fire and called on their 10-year-old brother to help extinguish it. The therapist hypothesizes that the family psychopathology resides within an executive subsystem comprised of the mother and a parental child, the 10-year-old brother. The youngster lighting fires is perceived as the family scapegoat. The goal of therapy is to move the parental child out of the conflict occurring between mother and daughter and return parenting functions to the mother. The conflict between mother and daughter is underlined by the target of the youngster's firesetting—her mother's bed. The therapist accomplishes the therapeutic goal by having the mother directly teach her daughter how to safely light and extinguish matches as a part of the process of therapy. In addition, to intensify the mother's parenting role, the therapist assigns the same task to be performed within the home. Hence, parenting responsibilities are returned to the mother, her 10-year-old son is freed to resume his role as a child, and her daughter begins to receive the necessary guidance and attention.

The daughter's firesetting behavior stopped after the first family therapy session and had not resurfaced at a two-year follow-up. The therapist maintained only an informal relationship with the family after termination of therapy. It is the therapist's opinion that the firesetting behavior was successfully eliminated when significant improvements became evident in the family's patterns of communication and interaction (Minuchin, 1974).

Madanes (1981) also reports success in eliminating firesetting behavior by utilizing controlled firestarting within the therapeutic setting. The youngster, a 10-year-old twin boy, the oldest of five children, had set numerous recent fires. In addition, the family had undergone several significant changes, including being abandoned by their father without financial support. The therapist viewed the youngster's firesetting as a vehicle for the expression of the mother's anger. Hence, the mother could punish and blame her youngster not only for firesetting, but for the rest of the family problems. The therapeutic goal was to shift the focus of the mother's experienced emotion from being angry at her son to being able to trust him. This was accomplished by the therapist first teaching the mother and her son, in the presence of the other family members, the correct way to safely ignite and put out matches. The therapist then instructed the mother to light a small fire and pretend to burn herself. The son was asked to put the fire out. In this way the youngster demonstrated to his mother that he knew how to correctly extinguish fires. The therapist asked the mother to repeat this procedure in the home. Hence, the son was able to playfully help his mother when she pretended to burn herself, and she in turn was able to trust him to put out the fire. The youngster's unsupervised firestarting stopped immediately and was not evident at a two-month follow-up. The mother continued in individual therapy for support and help with the remaining family problems.

Perhaps the most unique application of outpatient family psychotherapy to the treatment of firesetting is represented by the report of a case in which the youngster's firesetting behavior, the initial impetus for seeking help, never became the central focus of the therapeutic process (Eisler, 1974). The family entered treatment when their 14-year-old son, the oldest of three children, admitted to starting several large grass fires. The youngster had turned in the alarm when the fires had threatened several homes. He felt remorse for his behavior but could not explain it. The firesetting occurred about the same time that the father, after being absent from the family because he worked out of state, returned to live at home when local employment became available. In addition to significant changes within the family, the youngster was experiencing other difficult social circumstances such as poor school performance and confusion in relating to his peer group. His behavior also included a history of enuresis. The therapeutic goal was identified as helping the family members develop more effective methods of coping with the recently experienced series of crises, which included not only the son's firesetting, but the father's reentry into the family system. The therapist detected a dysfunctional family rule which prohibited the direct expression of dissatisfaction and, hence, viewed the firesetting as a nonverbal signal of family distress. The specific therapeutic technique employed was to ask family members to independently list their grievances about one another. Six two-hour family therapy sessions focused on such topics as how family members felt about the separation and return of father, the different roles assumed by everyone as a result of these changes, and the hopes they had for their future relationships with one another. The result of this intervention was a reopening of the communication channels between family members and candid discussions about their expectations of one another.

The therapist conducted four-month and one-year follow-ups with the family. The family reported no recurrence of the firesetting behavior. In addition, the son's enuresis had ceased and his peer relationships had improved significantly. There was no mention of changes in school performance. The therapist observed that there was a greater openness and livelier range of affect within the family system. This case suggests that in some instances firesetting may be a symptom representing a disturbance in social relationships, and an intervention strategy focused at adjusting the problematic relationships will be effective in eliminating the symptom of the distress, namely the firesetting behavior.

Although only three cases are reported utilizing outpatient family psychotherapy to treat pathological firesetting, they all demonstrate its effectiveness in adjusting patterns of communication, interaction, and the nature of role relationships, while at the same time eliminating the recurrence of firesetting within the family system. Hence, this particular therapeutic approach not only treats the presenting problem of firestarting, but it focuses on changing psychopathology associated with specific social circumstances. The application of family therapy techniques assumes that a major target of behavior change resides within the family. What remains unclear is the effectiveness of family therapy methods in adusting other behaviors, such as individual characteristics, which accompany firesetting and which also may be targets for change. However, Eisler (1974) does report the remission of enuresis, which is partially attributed to the success of the family psychotherapy intervention. In addition, the differential effectiveness of various types of family therapy techniques in treating a range of moderate to severe firesetting behaviors needs further examination. The application of outpatient family psychotherapy to the treatment of pathological firesetting shows remarkable promise, yet the short- and long-range effectiveness of this intervention strategy remains to be evaluated within the context of a controlled clinical experiment.

INPATIENT TREATMENT

There have been several attempts reported in the literature of treating youngsters presenting with pathological firesetting by removing them for a period of time from their living environment and utilizing an inpatient or residential milieu as the primary therapeutic approach (Siegel, 1957; Koret, 1973; Awad & Harrison, 1976; McGrath, Marshall, & Prior, 1979; Birchill, 1984). All of the inpatient treatment programs treating firesetting youngsters have several features in common. Each advocates an intensive and comprehensive evaluation and diagnosis process for youngsters and their families. In addition, all of the programs offer a multicomponent approach focused not only on eliminating firesetting behavior but on making significant changes in the psychological functioning of youngsters. Although most of the inpatient programs have similar treatment objectives for firesetting youngsters, they can be differentiated by the theoretical philosophies that underlie the organization and operation of these programs. It is these philosophical perspectives which dictate the specific types of treatment methods employed to help eliminate firesetting behavior and remediate the accompanying psychopathology.

There are two major types of theoretical philosophies influencing the inpatient treatment of firesetting youngsters. The first is a more traditional, psychodynamic approach where the treatment emphasis is on the nature of the therapeutic alliance formed between youngsters and the program staff. Both individual and family psychotherapy are techniques employed within the psychodynamic framework. The second theoretical approach is behavioral, where specific behaviors to be changed are identified and discrete interventions are designed to adjust these behaviors. The youngster is the primary focus of behavior therapy methods, although some of these inpatient programs also make an effort to work closely with the family. Inpatient programs representing both these theoretical philosophies will be described with respect to their methods of treating firesetting youngsters. The relative effectiveness of these inpatient programs in eliminating firestarting and resolving the accompanying psychopathology will be evaluated. Inpatient programs are a viable alternative treatment choice for helping firesetting youngsters.

Psychodynamic Programs

There have been three reports in the literature describing psychodynamically oriented inpatient treatment of youngsters presenting with pathological firesetting (Siegel, 1957; Koret, 1973; Awad & Harrison, 1976). These studies indicate a common set of features characterizing this type of treatment approach. These features include a philosophical thrust emphasizing the correlation between the emergence of firesetting behavior and the dynamic relationship between youngsters and their parents; the identification and interpretation of defense mechanisms that interfere with adequate emotional functioning; and the analysis of the psychological meaning of the firesetting behavior. These three features comprise the major themes of therapeutic treatment. A description will follow as to how these themes are worked through within the context of individual and family psychotherapy as well as within the therapeutic environment of the inpatient or residential treatment setting. The average length of this type of inpatient treatment is one year with a reported range of one to three years. The result of psychodynamically oriented inpatient treatment is the restoration of adequate ego functioning and the youngsters' adjustment to acceptable standards of social behavior.

Although each child-parent relationship contains its unique characteristics, the relationship between firesetting youngsters and their parents appears to have similar dynamic qualities which surface during the course of psychotherapy. The predominant feature is that parents tend to have highly ambivalent feelings about both being close and separating from their children. This is communicated to their youngsters on conscious as well as unconscious levels and leaves them confused and uncertain about their parental and social relationships. As a result of this ambiguity, these youngsters develop a symbiotic relationship with one or more parents, which is characterized by a high degree of dependence. In addition, there is an inadequate development of interpersonal skills, which results in social introversion and isolation. The goal of therapy is to establish a therapeutic alliance in which youngsters are encouraged to experience feelings of warmth and closeness while at the same time feeling trusting and secure enough to separate and develop meaningful relationships with staff, peers, and others within the therapeutic environment.

The second therapeutic theme, the identification and interpretation of defense mechanisms that interfere with adequate emotional functioning, occurs after the strength of the therapeutic alliance has been established between youngsters and their therapists. The defense mechanisms most typically employed by firesetting youngsters are suppression, repression, denial, and projection. Together these mechanisms work to prevent the expression of anger which is experienced within the relationship between youngsters and their parents. The premise is that youngsters cannot perceive or recognize their feelings of anger as well as modulate and express them in socially acceptable ways. The analysis and interpretation of this phenomenon is related by therapists to youngsters. An attempt is made to help youngsters recognize their angry feelings, understand why they feel the way they do, and teach them socially acceptable methods of expressing these feelings within their family and social network.

The final therapeutic theme, the analysis of the psychological meaning of firesetting behavior, is an interpretation that is offered to both youngsters and families within the context of psychotherapy. Two slightly different interpretations, each focused on firesetting behavior representing the unrealized expression of anger, dominate the literature. The first suggests that the firesetting behavior of youngsters is the acting out of the unexpressed aggression of their parent. The second indicates that firesetting is the unexpressed anger youngsters feel toward their parents. These interpretations of firesetting behavior are offered to youngsters and their families with the hope that understanding the meaning behind the behavior will prevent its recurrence.

All three cases report both the elimination of pathological firesetting and major adjustments in emotional functioning and social interactions at termination of psychodynamically oriented inpatient treatment. Unfortunately, none of the cases report any follow-up information; therefore, the long-term effects of this type of treatment cannot be evaluated. In addition, because the duration of these programs averages one year, a relatively long period of time, the question arises as to whether there is a more cost-effective treatment option. It is difficult to begin to address this question both because there is no information concerning the long-term effectiveness of these programs, and in general, the longer the treatment, the higher the cost. If significant improvement occurs in the lives of firesetting youngsters and their families, given the relative time commitment and cost of psychodynamically oriented inpatient treatment, then the benefits may outweigh the liabilities of this therapeutic approach.

Behavioral Programs

The literature contains two reports of the inpatient programs utilizing behavior therapy methods to treat pathological firesetting (McGrath, Marshall, & Prior, 1979; Birchill, 1984). Both inpatient treatment approaches report the application of intensive, short-term (4- to 12-week) behavior therapy techniques that successfully stop firestarting and adjust additional behaviors which represent accompanying psychopathology.

The suppression of an 11-year-old youngster's firesetting was achieved by designing an intervention program based on four major hypotheses (McGrath et al., 1979). The first two hypotheses, that firesetting occurs following difficulties in coping with stress and in relating to peers, was treated by social skills training. Difficult social interactions were presented to the youngster, and the inappropriate responses were videotaped and played back. Through role playing, modeling, and rehearsal, the youngster learned the correct social responses. The third hypothesis, that firesetting is reinforcing, was treated by negative practice or overcorrection. This procedure, also utilized in outpatient treatment (Kolko, 1983), requires youngsters to ignite controlled fires in a metal container, extinguish them, and clean the resulting burn and scorch marks. Throughout this process the youngster is told to repeat basic fire safety rules such as "fires are dangerous." The fourth hypothesis, that the dangers of firesetting are not realized, was treated with covert sensitization and a fire safety project. The covert sensitization procedure involved the youngster first hearing a tape depicting himself in the middle of a dangerous firestarting incident. Following this anxiety-inducing situation, a relief scene is played in which the youngster manages to control his impulse to firestart and finds an alternative way of coping with the experienced stress. The fire safety project consisted of a tour of the local fire department, a visit to a recently burned building, viewing three fire safety films, reading several chapters of a book on fire safety and answering questions, and presenting a fire safety project for evaluation by the fire chief. This four-component intervention program was implemented in a residential detention home.

During the 12-week course of inpatient treatment, three collateral behaviors, in addition to firestarting, were identified as targets for change. They were the youngster's ability to cope with stress, the frequency with which he engaged in social activities, and his participation in additional positive behaviors such as volunteering and cooperation. There were significant improvements in two of the three targeted areas—coping with stress and engaging in social activities. In addition, it was reported that a smooth transition was made from the residential treatment center to a foster home, where his parents reported good progress in such age-appropriate behaviors as making friends, participating in family activities, and school achievements. A two-year follow-up indicated no recurrence of firesetting and adequate adjustment to both home and school environments.

The second inpatient program based on a behavior approach also utilizes a combination of individual and family psychotherapy to remediate pathological firesetting (Birchill, 1984). The program is comprised of three major phases. In the first phase, the youngster enters a hospital setting for a four-week stay. During this time, the youngster participates in a series of exercises in which a choice must be made between toys and firestarting materials. These exercises are observed by a therapist out of view of the youngster. If the youngster chooses matches and lighters as opposed to other, non-fire-related toys, the therapist intervenes and debriefs the youngster. The debriefing process focuses on helping the youngster realize the experienced emotions associated with choosing firestarting materials. The result of these exercises is that the youngster is left with a mild aversion to firestarting materials.

The second and third phases of this behaviorally oriented inpatient program employ intensive family psychotherapy to adjust the patterns of communication and interaction within the family system. During the second phase, family psychotherapy is initiated while the youngster remains in the hospital setting. The premise underlying the introduction of family psychotherapy is that certain environmental conditions exist within the family which are associated with the emergence of the youngster's firesetting behavior. These conditions must be changed if the youngster is to return to the family at discharge. Therefore, significant emphasis is placed on improving the family environment so that the youngster will return to a supportive, trusting family system. The third phase of the program focuses on the youngster's reintegration within the family as well as the community. Outpatient family psychotherapy is employed, and an intensive effort is undertaken to provide special community support services such as schooling and structured activities for the youngster. At the end of the third phase, it is expected that the youngster has returned to the family and social environment, is not participating in firestarting activities, and is functioning adequately within the interpersonal and social milieu.

Although no formal studies appear in the literature regarding the short- or long-term success of this program, one newspaper account reveals that of the 100 youngsters treated in a two-year period, no reported relapses have occurred (Abrams, 1985). It is unclear from this information whether an active attempt was initiated to obtain follow-up data on these youngsters and their families after treatment was terminated. In addition, there do not appear to be any data regarding the effectiveness of this inpatient treatment approach in remediating the psychopathology accompanying firesetting behavior, such as those patterns of family communication and interaction which are the focus of the application of family psychotherapy. Therefore, although preliminary information suggests that this behaviorally oriented inpatient program is successful in treating pathological firesetting, before such a claim can be substantiated, more data are necessary regarding exactly what components of the program are effective in treating what aspects of both the youngster's behavior and the family environment.

In general, all of the reported attempts to treat pathological firesetting with an inpatient approach, whether it is psychodynamic or behavior therapy, lack empirical data regarding the short- and long-term effectiveness of eliminating firesetting and the psychopathology that accompanies the behavior problem. There are many questions to be addressed regarding the application of inpatient treatment to pathological firesetting. For example, are inpatient programs appropriate only for those severe cases of youngsters presenting with firesetting behavior who must be removed from the community because they are a danger to themselves and others? Or is there a specific range of firesetting youngsters for which inpatient treatment is most appropriate? In addition, what is the relative cost-effectiveness of inpatient treatment? Do the benefits of these programs in treating pathological firesetting outweigh the relatively high costs of inpatient and residential treatment programs? It is critical that future research efforts focus on these questions so that an understanding can be achieved regarding the appropriate application of various types of inpatient treatment.

DETERMINING PSYCHOTHERAPY OF CHOICE

Throughout the chapter, the intervention strategies applied to the treatment of pathological firesetting have been presented according to their predominant psychotherapy methods. Outpatient treatment approaches, including cognitive-emotion, behavior, and family psychotherapy, have been described and the evidence reviewed regarding their relative effectiveness. Inpatient treatment programs, utilizing psychodynamic and behavior therapy methods, also have been detailed, and the available information has been outlined concerning their effectiveness. Given these treatment options, how is the determination made regarding the best fit between youngsters and families presenting with pathological firesetting and the psychotherapy of choice?

The answer to this question would be greatly facilitated if adequate research information existed on the relative effectiveness of various clinical treatment approaches in eliminating firesetting behavior and changing the targeted behaviors associated with the accompanying psychopathology. Given the lack of scientific data to contribute to the clinical decision-making process, standards must be developed and utilized as guidelines for determining which psychotherapy methods are best suited for specific types of youngsters and families presenting with pathological firesetting. There are two major classes of treatment decisions. The first is the decision of whether the most appropriate type of treatment is outpatient or inpatient. Once the decision has been made regarding outpatient or inpatient treatment, then the next step is to determine the type of treatment orientation. Treatment programs can apply psychodynamic or behavior philosophies, and they can employ a variety of techniques including individual and family psychotherapy. The primary goal is to achieve an optimal match between the identified clinical problems of youngsters and their families and the most beneficial intervention strategy.

Outpatient or Inpatient Treatment

The first step in the clinical decision-making process is determining whether outpatient or inpatient treatment is the most appropriate intervention for youngsters presenting with pathological firesetting. In general, a thorough clinical evaluation of youngsters and their families will yield an adequate assessment of the severity of firesetting behavior as well as identification of those behaviors representing the accompanying psychopathology targeted for change. The outcome of this clinical assessment must be matched with those features characterizing the treatment objectives of outpatient and inpatient programs. However, in addition to determining this critical match, an assessment must be made of the potential for future firestarting behavior. If this assessment indicates a high risk, then the decision of inpatient treatment must be given a high priority.

The major indication of high risk for future involvement in firesetting behavior is the admission by youngsters of a plan for their next firestart. The more well thought out and articulated the plan, the more probable the involvement in future firesetting incidents. Hence, during the clinical interview, youngsters must be questioned as to whether they have thought about their next firestart. If the answer is yes, further questions must be asked regarding their plan. The more detailed the plan, the higher the risk for future involvement. If the plan includes such components as a specific target, time, place, method of ignition, and motivation, then clinical consideration must be given as to whether youngsters are a serious threat or danger to themselves or others. If there appears to be an insistent determination to persist in firesetting and no agreement can be reached regarding a commitment not to participate in firesetting, then an inpatient treatment program is the best clinical option.

Some youngsters may admit to future firesetting plans and will even specify details such as target and motivation, but they are usually willing to agree to stop firesetting. In addition, they often will agree that if they experience the impulse of wanting to set a fire, they first will talk about it with either a parent, an adult in whom they can confide, or a clinician. For these youngsters, an inpatient treatment program still may be appropriate, but may not be mandatory. If youngsters agree to stop future firesetting, but their clinical assessment indicates severe psychopathology with multiple behaviors targeted for change, then inpatient treatment must be given strong consideration. Given the existence of severe psychopathology accompanying firesetting behavior, inpatient intervention may be the treatment of choice in effecting significant change in the cognitive and emotional functioning of youngsters as well as adjustments in family environment and interpersonal and social behavior.

If youngsters are not a threat or danger to themselves or others and the existence of severe psychopathology can be ruled out, then outpatient intervention becomes a viable treatment option. Several reports indicate the relative success of outpatient procedures in eliminating firestarting behavior in remarkably short periods of time. In addition, depending on the outpatient psychotherapy method, these reports also indicate success in changing those behaviors representing associated psychopathology. If the probability is low that youngsters will exhibit firesetting behavior in the near future, and the probability is high that the accompanying psychopathology can be adjusted, then the chances are excellent for a successful outpatient treatment intervention.

Choosing Psychotherapy Methods

The choice of psychotherapy method, whether an outpatient or inpatient intervention strategy is employed, depends on the optimal match between the specific clinical behaviors targeted for change and the major features characterizing the philosophy, objectives, methods, and effectiveness of treatment programs. This optimal match is achieved by choosing the best fit between what a thorough clinical assessment indicates as the major behaviors targeted for change and the treatment objectives and the modalities utilized to achieve them. Clinicians must evaluate each case according to the critical elements comprising the evaluation and decide, given information on the major features characterizing psychotherapy methods, the best course of therapeutic intervention.

Table 6.1 summarizes the major features describing the various types of outpatient psychotherapy methods. The major features are the psychological philosophy or theory underlying the treatment approach, the modalities employed, the type of direct interventions utilized to eliminate firestarting, the techniques applied to change the targeted behaviors reflecting accompanying psychopathology, and the benefits and liabilities of each of the therapeutic approaches. All psychotherapy methods reportedly are successful in eliminating firestarting. However, they differ dramatically in their philosophical approach to treatment and, hence, their types and targets of inter-

Table 6.1
Predominant Features Characterizing Outpatient Treatment Programs

Treatment Method	Philosophy	Modalities	Firesetting Intervention	Targets of Behavior Change	Benefits	Liabilities
Cognitive-emotion	Psychodynamic/cognitive-emotion	Individual psychotherapy with occasional participation by parents	"Graphing technique"	Changes in emotional functioning as related to firesetting; learning to cope with stressful life events; learning appropriate methods of expressing anger	Eliminates firestarting; adjusts emotional and cognitive functioning	No evidence indicating success with older youngsters exhibiting longstanding firesetting histories
Behavior therapy	Social learning theory	Individual, parent, and family therapy	Punishment, reinforcement, negative practice/satiation, and operantly structured fantasies	Replacing aggressive behaviors with more socially compliant behaviors	Eliminates firestarting; a variety of behavior therapy methods are successful in stopping firesetting	Individual behavior techniques only applied to single cases; does not typically treat accompanying psychopathology, although some methods can be applied
Family psychotherapy	Psychodynamic	Family psychotherapy	Corrective practice	Changes in family role relationships and patterns of communication and interaction	Eliminates firestarting; pervasive changes in family environment	Unclear how effective in correcting accompanying psychopathology related to individual characteristics

ventions. If a youngster's clinical assessment indicates outpatient treatment as an appropriate option, then the next step is to determine the match between what the clinical evaluation identifies as the targets of behavior change and the psychotherapy of choice. If the clinical evaluation suggests that treatment primarily focus on adjusting the emotional functioning of the youngster, then cognitive-emotion psychotherapy may be the optimal match. If the clinical assessment identifies aggressive behaviors accompanying firestarting, then the application of behavior therapy techniques may be more appropriate. If the clinical evaluation uncovers significant family psychopathology related to the firesetting behavior, then family psychotherapy may be the best treatment option. Because all of these approaches effectively stop pathological firestarting, the critical decision in selecting the optimal intervention strategy is in choosing the psychotherapy method that will successfully adjust the predominant targets of behavior change associated with the accompanying psychopathology.

Table 6.2 outlines the major features detailing inpatient treatment programs. Like the outpatient treatment methods, the major features include the psychological philosophy underlying the treatment approach, the modalities employed, the type of intervention directed at eliminating firestarting, the techniques applied to change the targeted behavior associated with the psychopathology, and the benefits and liabilities of the therapeutic approaches. If the youngster's current firestarting behavior is a serious threat to himself or others or the accompanying psychopathology is severe, then inpatient treatment is a high priority. There are two inpatient treatment approaches, both of which are successful in eliminating firestarting behavior. The psychodynamic method emphasizes major changes in emotional functioning and the dynamic relationship between the youngster and parents. The behavior method creates significant changes in coping with stress, social interaction with peers, and the structure of the family environment. Each inpatient approach supports adjustments in many areas, with the psychodynamic method emphasizing changes in individual functioning and the behavior method encouraging changes in social functioning. Once the clinical assessment identifies the predominant targets of

Table 6.2
Predominant Features Characterizing Inpatient Treatment Programs

Treatment Method	Philosophy	Modalities	Firesetting Intervention	Targets of Behavior Change	Benefits	Liabilities
Psychodynamic	Psychoanalytic	Individual, family, and milieu psychotherapy	Analysis of the psychological meaning of the firesetting behavior	Changes in emotional functioning with an emphasis on adjusting defense mechanisms; changes in relationship between youngster and parents	Eliminates firestarting; major changes in emotional and social interaction	Long-term (average one year) treatment coupled with high cost
Behavior	Social learning theory	Individual and family therapy	Social skills training, negative practice/overcorrection, covert sensitization, and fire safety project	Acquires skills in coping with stress and changes methods of social interaction with peers; restructures family environment	Eliminates firestarting; places youngster back into a healthy family system	Unclear how effective specific components of treatment work with different types of youngsters

behavior change, then the inpatient program must be selected on the basis of its ability to sustain significant changes in the indicated psychopathology.

The effectiveness of the psychotherapy depends on maximizing the fit between the specific information regarding firesetting behavior and the accompanying psychopathology contained in the clinical assessment and the mechanisms and targets of change employed by particular treatment programs. The result of this decision-making process depends on the appropriate application of clinical data. The optimal match between the clinical needs of youngsters and their families and the focus of psychotherapy programs represents the best possible effort at effecting a good prognosis and a positive treatment outcome.

SUMMARY

The major outpatient and inpatient psychotherapy methods are described with respect to their underlying psychological philosophy, the treatment modalities employed, and the intervention techniques directed at both eliminating pathological firesetting and changing those targeted behaviors reflecting accompanying psychopathology. Regardless of the specific psychotherapies utilized, and whether they are applied in an outpatient or inpatient treatment setting, they all report success in stopping firestarting behavior. The distinguishing features of these psychotherapies rests with their relative effectiveness in adjusting the various target behaviors representing the psychopathology that accompanies pathological firesetting.

The selection of the best intervention strategy depends on maximizing the fit between the clinical assessment data describing the nature and severity of the firesetting behavior and the associated psychopathology and the philosophy, modalities, and targets of behavior change emphasized by the treatment programs. If clinical assessments reveal youngsters who are a significant danger to themselves or others and multiple targets of behavior change indicate severe psychopathology, then inpatient treatment must be strongly considered. The two approaches to inpatient treatment are psychodynamic and behavior therapy. The psychodynamic method employs individual, family, and milieu psychotherapy in an effort to effect major reconstructive changes in emotional functioning and family environment. The behavior method applies individual and family therapy to adjust skills in coping with stress, improving social interaction with peers, and reshaping the family system.

If the need for inpatient treatment is ruled out, then outpatient psychotherapy becomes a viable intervention option. The three predominant outpatient treatment methods are cognitive-emotion, behavior, and family psychotherapy. In general, the cognitive-emotion approach uses individual psychotherapy to effect changes in emotional functioning; behavior therapy combines individual and family therapy to replace firestarting with more socially acceptable behaviors; and family psychotherapy works within the family structure to adjust role relationships and patterns of communication and interaction.

Clinical evidence suggests that all of these psychotherapies are effective in eliminating pathological firesetting and improving the accompanying psychopathology. However, there are few empirical studies to support these claims of clinical success. Until appropriate scientific standards of evaluation, such as adequate research designs and the application of reliable and valid outcome measures, are utilized to examine the quantitative effectiveness of these psychotherapies, the qualitative, clinical evidence must stand by itself as support for the success of these intervention strategies.

7

Pathological Firesetting and Community Intervention

Psychotherapies do not stand alone as the unique intervention option in the treatment of pathological firesetting. Recently there has been a concentrated effort to develop community-based intervention programs to identify, educate, counsel, and establish referral linkages for youngsters and families presenting with the problem of recurrent, nonproductive firestarting. There are two major types of community programs. The first are those whose primary goal is prevention. They operate under the assumption that fire interest is a curiosity that naturally occurs in children. Therefore, an effort must be made to educate all youngsters in fire safety. These programs generally are offered in school settings; however, there is one national effort conducted both through the media, utilizing television and newsletters, and through the schools. The second type of community program also includes a prevention orientation; however, there is an additional emphasis on early identification and intervention with those youngsters who evidence patterns of repeated firestarting. These programs usually are maintained by fire departments, which in turn network with a variety of other community agencies.

Examples of each of these two types of community intervention strategies are described, their effectiveness in eliminating nonproductive firestarting is examined, and the attention they pay to the psychopathology accompanying recurrent firesetting is evaluated. A model is proposed that describes an integrated framework comprised of community intervention and psychotherapy which is designed to reduce the incidence of unsupervised fireplay, nonproductive firestarting, and juvenile-related arson. This proactive model raises significant questions about the relative effectiveness of intervention strategies and contributes to the identification of those relevant dimensions which must be the focus of future clinical research.

PREVENTION

Although their availability varies from community to community, there are a number of prevention programs which are designed to teach children how to be fire-safe. These programs operate under the assumption that fire interest is a naturally occurring curiosity which emerges in most youngsters as early as three years of age. Like many predispositions, such as water play, children must learn the circumstances under which their fire interest can be pursued under safe and instructional conditions. Becoming fire-safe is viewed as a survival skill, much like becoming water-safe; therefore, fire safety skills, like learning to swim, must be taught to youngsters. If fire safety skills are not acquired, then the risk is high that the predisposition toward fire interest could result in fire-risk behaviors such as fireplay or nonproductive firestarts. Hence, prevention programs acknowledge youngsters' natural curiosity about fire and discourage the emergence of fire-risk behaviors by teaching the rules of fire safety.

Four major approaches to fire prevention will be described because they represent the spectrum of age-appropriate methods currently applied in teaching fire safety and prevention. These programs include the Children's Television Workshop's (CTW's) Fire Safety Project (1982), the National Fire Protection Agency's (NFPA's) Learn Not to Burn Firesafety Curriculum (1979), the St.

Paul, Minnesota Fire Education Program (National Committee on Property Insurance, 1984) and an experimental program focused on the Training of Emergency Fire Escape Procedures (Jones, Kazdin, & Haney, 1981). Although all of these programs primarily operate in school settings, they differ dramatically in the type of educational methods they employ. These contrasts are largely due to the different age groups targeted by these programs. The effectiveness of these prevention programs will be reviewed in terms of their success in teaching youngsters how to be fire-safe as well as deterring fireplay and nonproductive firestarting behavior. Because prevention programs generally are aimed at educating the "normal" population of youngsters, questions will be addressed regarding whether these programs can be utilized to help youngsters presenting with pathological firesetting and the accompanying behavior problems.

Children's Television Workshop's Fire Safety Project

The Children's Television Workshop's (CTW's) Fire Safety Project, produced by its Community Education Services Division (1982), is a unique nationally focused program teaching preschool youngsters fire prevention. This program supports recent studies (Kafry, Block, & Block, 1981) indicating that fire interest emerges in most children as early as the age of three. Hence, educational efforts aimed at preschoolers, who are in the initial stages of expressing their curiosity about fire, are likely to be effective in channeling their interest in a productive and positive manner. The CTW's Fire Safety Project reaches youngsters at the critical age of their initial fire awareness and teaches them appropriate attitudes toward fire as well as basic fire safety rules.

The methods and mediums of communication utilized by the CTW's Fire Safety Project deserve special attention. One of the most popular teaching methods is the application of Sesame Street characters (already popular with preschoolers) as the primary communicators of fire safety lessons. Characters like "Bert" and "Ernie," two very different fellows but very best friends, are utilized to teach children about fire drills, firefighters, and firefighter training. Because these Sesame Street characters already have entered the lives of many preschoolers through a variety of different types of exposure, they become effective communicators of important messages to youngsters.

The CTW's Fire Safety Project employs multiple mediums of communicating fire safety information. The primary method of teaching fire prevention to preschoolers is through short, single-topic vignettes presented as a regular part of the television programming of Sesame Street. Initial research indicated that although some fire safety lessons were appropriate for television viewing, the majority of the material needed to be demonstrated directly to youngsters (CTW, 1982). Hence, CTW designed materials and seminars to teach children in a preschool setting. In addition, the CTW produces a Fire Safety Newsletter available to nursery and preschools. The application of three diverse mediums of communication—television, teaching materials in small seminar settings, and the Fire Safety Newsletter—coupled with the use of Sesame Street characters as communicators offers an intensive and consistent approach to teaching fire safety to youngsters at the critical preschool age.

Preliminary evaluation research on the CTW's Fire Safety Project indicated that a broader and more intensive approach to teaching children fire safety, in addition to television presentations, was likely to be more effective in directly reaching preschoolers (CTW, 1982). As a result, the CTW expanded their efforts to teach fire prevention by including the use of multiple methods such as the development of materials to be used in small seminars with youngsters and the distribution of newsletters to preschools and nursery schools. Because the CTW adopted this type of broad and intensive approach, it becomes difficult to measure or quantify the effectiveness of such a program. Case reports by teachers who have employed the CTW's educational materials with their preschoolers have indicated that youngsters are successful in learning fire safety rules. Although there has not been an attempt to evaluate the entire effort of the CTW's Fire Safety Project, which would require a long-term and somewhat costly evaluation research project, preliminary studies and case reports suggest that preschoolers are capable of learning fire prevention.

Given what is known about the development of fire interest in youngsters, age-appropriate instruction aimed at teaching the rules of fire prevention to preschool children appears to be an excellent first step in promoting early fire-safe behavior skills. Whether the early introduction of such skills will deter the development of fire-risk behavior, such as fireplay and nonproductive firestarting, remains open to future question and research.

Learn Not to Burn

The National Fire Protection Agency's (NFPA's) Learn Not to Burn Firesafety Curriculum was developed to meet the growing demand to educate elementary-school-aged youngsters about the rules and behaviors of fire prevention. The NFPA mounted an extensive project designed to develop and field-test an elaborate package of materials for teachers to utilize in instructing their students in fire safety. The goals, objectives, and materials comprising the Learn Not to Burn program will be outlined for the purpose of describing a comprehensive set of information regarding key fire safety behaviors to be taught to "normal" elementary-school children.

The goals of the Learn Not to Burn Firesafety Curriculum are to reduce the number of deaths and injuries caused by fires and to reduce the numbers of fires and the extent of property loss. To accomplish these goals, a curriculum, accompanied by teacher materials, was developed to be taught sequentially throughout the elementary-school grades. The curriculum is organized around a set of 25 key firesafety behaviors which are divided into three categories: protection, prevention, and persuasion. It is recommended by the NFPA that youngsters demonstrate competence in all 25 behaviors by completion of elementary school.

The three categories of firesafety behaviors—protection, prevention, and persuasion—are prioritized in terms of their importance. The first priority, protection, involves those behaviors youngsters must learn in case of fire. Some examples of protection behaviors include participating in school fire drills, performing the stop, drop, and roll procedure, and developing a home fire escape plan. The second priority, prevention, is comprised of those behaviors which will deter injury before fire occurs. Some examples of prevention behaviors include how to use matches safely, how to store flammable liquids, and how to identify and remove electrical hazards. The third priority, persuasion, focuses on those behaviors which encourage others to become aware of fire safety and prevention activities. Some examples of persuasion behaviors involve teaching others about smoke detectors, making sure others properly maintain their electrical equipment, and helping others install electrical outlet covers. Once youngsters learn these 25 key firesafety behaviors, it is expected that they will exhibit a continuing awareness and responsibility in fire prevention.

The primary contribution of the Learn Not to Burn program is an impressive package of materials for teachers which is designed to help them implement the curriculum within their classroom setting. These materials include documentation on the general design of the curriculum, 25 curriculum cards showing teachers how to help their students attain competence in each of the key firesafety behaviors, evaluation instruments designed to measure achievement in fire prevention and satisfaction with the program, fire prevention information for teachers to disseminate to their students, and information regarding the availability of additional teaching aids. The elements comprising the Learn Not to Burn Curriculum are a thorough compilation of instructional materials designed to help teach firesafety behaviors to elementary school children.

From the research and development phase to the dissemination of the Learn Not to Burn Curriculum, the NFPA has paid close attention to evaluating the effectiveness of this instructional program. They report conducting an evaluation research study, utilizing control groups, to assess the effectiveness of implementing Learn Not to Burn in seven urban and suburban sites involving 4,000 students and 200 teachers. The impact of the curriculum on both knowledge and practice was found to be significantly better for those youngsters participating in the learning program. Parents reported being highly satisfied with their childrens' involvement in Learn Not to Burn. In addition, teachers gave highly positive ratings to the curriculum in terms of its value and its usefulness. Based on these evaluation data, the NFPA published the Learn Not to Burn Firesafety Curriculum and began marketing it on a nationwide basis. The program is currently implemented in numerous school systems throughout the country.

Although initial reception and acceptance of Learn Not to Burn are extremely favorable, as yet there are no follow-up data regarding the long-term effectiveness of this type of prevention program in meeting its stated goals of reducing the number of fires, extent of property damage, and fire-related deaths and injuries. It would be useful to compare fire incidence rates, especially those fires related to juvenile firestarts, in communities where Learn Not to Burn has been implemented versus similar demographic communities where there has been no participation in a systematic fire education program. Moreover, little is known about the impact of this educational effort in discouraging participation in nonproductive fireplay and firestarting behavior. For example, it would be useful to know whether youngsters exposed to the Learn Not to Burn Firesafety Curriculum are less likely than their unexposed counterparts to engage in the actual behaviors of unsupervised fireplay or firesetting. If this is the case, then this type of educational program will not only be helpful to the ''normal'' population of elementary-school children, but it may be useful in preventing future, nonproductive firestarting behaviors in those youngsters who already exhibit patterns of pathological firesetting.

The St. Paul Fire Education Program

The St. Paul, Minnesota, Insurance Companies have sponsored a unique educational program aimed at teaching fire prevention to eighth and ninth graders, who in turn utilize their newly acquired information and skills to instruct fourth and fifth graders on the topic of fire safety. The major advantage of this approach to prevention is that it educates an often neglected age group (adolescents) who frequently are at high risk for involvement in nonproductive firestarts. In addition, it builds their self-confidence by introducing them not only to new information, but to new roles as teachers. In this way, two different age groups of youngsters learn the rules of fire safety.

The St. Paul Fire Education Curriculum not only teaches fire prevention, but it also emphasizes general topics in crime prevention. The curriculum is designed to provide youngsters with information on arson, vandalism, property crimes, law enforcement, and the juvenile justice system. The curriculum consists of lesson plans on each of these topics for each day of a 12-week course. There are exercises that teach the eighth and ninth graders how to apply their newly acquired knowledge of fire safety and how to prepare themselves for their forthcoming roles as student teachers. In addition to a documented curriculum, the St. Paul Fire Education Program encourages the participation of law enforcement, probation, and juvenile justice personnel in helping to teach youngsters not only the principles of crime prevention, but also a sense of responsibility toward their school and their community.

The St. Paul Fire Education Curriculum contains an evaluation component which requires participants to provide feedback on the effectiveness of the program. Case reports suggest that both eighth- and ninth-grade as well as elementary teachers are highly satisfied with the program. In addition, both eighth- and ninth-grade and elementary students report not only being highly satisfied with their participation in the program, but they indicate learning a significant amount of new material related to fire prevention. Although these case reports are highly favorable, there does not appear to be a systematic effort underway to test the relative effectiveness of this program in deterring the occurrence of antisocial behavior in youngsters. If this prevention approach demonstrated a significant impact in reducing the incidence of youthful criminal activity in those communities which have implemented the program, then arguments could be made for its application in a number of different settings to help prevent youngsters from participating in nonproductive firestarting as well as other antisocial behaviors.

Training Emergency Fire Escape Procedures

Because a significant proportion of fires occur in the home each year (Hartford Insurance Group, undated), and because loss of life is frequently associated with fires in the home at night (US Department of Commerce, 1978), a multifaceted behavioral program was designed to teach youngsters the appropriate responses to fire emergencies occurring in their home at night (Jones, Kazdin, & Haney, 1981). The goal of this program is to decrease the risk of youngsters being severely burned or overcome by smoke or panicking, and to increase the probability of youngsters reaching safety once they discover themselves in a fire emergency situation.

The training program consists of simulating nine home fire emergency situations. These situations and the correct responses to these fire emergencies were developed in collaboration with national fire agencies and local fire officials in the Pittsburgh area. An example follows of one of these fire emergency situations and the sequence of correct responses that must be demonstrated successfully by youngsters.

Situation	*Responding Requirements*
The experimenter told the children that they were coughing and their eyes were burning and they could not leave their bedroom through the window if they needed to.	Correct responding required the children to: a) slide to the edge of the bed b) roll out of bed c) get in a crawl position d) crawl and get the rug e) push the rug in the crack f) crawl to the window g) open the window h) yell and signal for help

Training youngsters to respond correctly to all nine simulated home fire emergency situations involves 13 discrete steps. The trainers provide a (1) verbal and (2) practice review of the correct responses; (3) they model the correct behaviors; they deliver feedback for (4) correct and (5) incorrect responses and (6) praise for correct performance; (7) they review new material; (8) they involve other trainers in the process; (9) they allow youngsters to reward themselves for a good performance; they allow children to (10) select and (11) place their rewards in a special place; (12) they acknowledge the youngsters' correct responses; and (13) they give appropriate verbal and nonverbal cues for the desired behaviors. The average training time was slightly under five days of two 20-minute sessions per day to obtain the expected behaviors from youngsters on all nine simulated fire emergency situations.

To assess the effectiveness of the training program, five children, ranging in age from eight to nine years, with normal to low-normal levels of intelligence and who scored near zero levels of performance on an initial fire safety skills screening assessment, were selected to learn the nine home fire emergency situations. All of these children lived in a section of Pittsburgh, as reported by the Fire Department, that ranked twelfth of 42 neighborhoods (seventy-second percentile) in the number of annual fire alarms.

Baseline data indicated that all five of these youngsters were unable to respond correctly in home fire emergencies occurring at night before their participation in the training program. Both verbal and performance measures showed a significant improvement in their fire safety skills after their participation in the training program. A two-week assessment indicated that these youngsters were able to maintain their newly acquired fire escape and safety skills. Although the success of training emergency fire escape procedures is documented for these youngsters, it is unclear how effective this program is when implemented on a larger scale with a variety of other types of children. In addition, little is known about the impact of this prevention technique in reducing the actual number of injuries and deaths due to home fire emergencies. Therefore, although this program to train youngsters in fire escape skills appears to address an important aspect of fire safety and prevention, and preliminary data suggest its success, its long-term impact and wider application remain open for further investigation.

The Application of Prevention Programs

Prevention programs are part of a general strategy designed to teach youngsters how to be fire-safe. The appropriate application of various prevention programs in teaching age-appropriate fire safety skills must take into account such factors as their targeted age group, their objectives or outcomes, their teaching methods, and the information available on the effectiveness of their particular approach to prevention. Table 7.1 summarizes the specific types of prevention programs according to these four major factors. Effective implementation of any prevention program depends on careful consideration of the composite of these factors.

All of the prevention programs share one common feature; that is, they aim at educating the ''normal'' population of school-age youngsters. Beyond this, they all vary with respect to the four major factors—age, objectives, teaching methods, and effectiveness—which characterize their organization and operation. Because the prevention programs target different age groups, the educational needs and developmental skills vary for each of the programs. Consequently, the objectives, anticipated program outcomes, and teaching methods are different depending on the age of the student population. For example, the CTW's Fire Safety Project is targeted for preschoolers, emphasizes the acquisition of basic fire safety knowledge, including information about fire drills and firefighters, and utilizes popular Sesame Street characters, such as Bert and Ernie, as communicators of the rules of fire prevention. In contrast, the St. Paul Fire Education Program aims at teaching eighth and ninth graders not only about fire safety but about the various aspects of crime prevention, instructs these youngsters in how to present their newly acquired information to fourth and fifth graders, and thereby educates two different age groups by employing two different teaching methods. Each of these prevention programs makes a unique and significant contribution to fire safety education.

Most of these prevention programs utilize an internal evaluation system designed to measure the amount of information acquired by their student audience as well as how satisfied students, teachers, and parents are with the implementation of the programs. These are appropriate assessment methods for gauging the short-term impact of these prevention programs. However, there have been virtually no systematic efforts to measure the relative impact of these prevention strategies on long-term outcomes such as the reduction of juvenile-related fires and associated property damage or the decrease in the number of youthful injuries and deaths caused by fires. In addition, little is known about the effect of these programs in discouraging unsupervised fireplay and participation

Table 7.1
The Major Factors Characterizing Fire Prevention Programs

Program	Target Age	Goals/Objectives	Teaching Methods	Effectiveness
The Children's Television Workshop's Fire Safety Project	Preschool	To teach age-appropriate information related to fire drills, firefighters, and firefighter training	Popular children's characters utilized as communicators Multiple methods including television, small seminars, and newsletters	Case reports by teachers show positive acceptance of materials and their application
Learn Not to Burn Curriculum	Elementary school	To reduce fire incidence and the number of related injuries and deaths; to teach three basic areas of fire safety—protection, prevention, and persuasion—which are comprised of 25 fire-competent behaviors	A comprehensive curriculum guide for teachers, which includes written educational materials for children	Positive ratings by teachers and parents and significant gains in fire safety knowledge by youngsters during the curriculum's research-and-development phase
The St. Paul Fire Education Program	Eighth and ninth grades Fourth and fifth grades	To teach fire safety and crime prevention to eighth and ninth graders and to build their self-confidence by utilizing them as role models and teachers of fire prevention to fourth and fifth graders	A documented curriculum and lesson plans plus participation of community officials representing law enforcement and juvenile justice in teaching activities	Student case reports claiming high satisfaction and significant gains in knowledge of fire safety and prevention
Training Emergency Fire Escape Procedures	Elementary school	To train youngsters to respond correctly to home fire emergency situations	Nine simulated home fire emergency situations with accompanying correct responses	Five children showed significant improvement in responding behavior after exposure to training

in nonproductive firestarting for the "normal" population of youngsters as well as for those who have histories of firesetting behavior. There is some evidence suggesting that communities implementing some type of fire prevention program can expect a 90% to 100% drop in nonrecurring firesetting behavior or recidivism rates (National Committee on Property Insurance, 1984). More specific evaluative data are needed to assess which youngsters in which communities would benefit most from what types of fire safety and prevention programs. Understanding the conditions under which these programs are successful and documenting the exact nature of their effectiveness will help substantially in their application to teach youngsters fire-safe behaviors and to prevent them from participating in nonproductive firesetting.

EARLY IDENTIFICATION AND INTERVENTION

Prevention programs employed in school settings represent one community intervention approach to reducing nonproductive firestarting by educating youngsters about the rules of fire safety. It has been suggested that these programs are most effective in teaching the "normal" population of school-age youngsters how to be fire-safe. Little is known about their impact on the behavior of those youngsters who already have expressed their interest in fire by participating in fireplay or nonproductive firesetting. Recently, fire service organizations have begun to develop and implement methods not only aimed at early identification of youngsters with histories of firestarting behavior, but directed at providing early intervention programs to eliminate firesetting as well as the accompanying behavioral problems.

The majority of these community intervention programs are operated by local fire departments. These grass-roots efforts to help youngsters often are supported by national fire service organizations, which provide technical and financial assistance to maintain the operation of these intervention programs. Three programs will be presented which represent the three predominant community intervention strategies currently employed by fire departments. They are the Fire-Related Youth Program (FRY Program) of the

Rochester, New York Fire Department (Cole, Laurentis, McAndrews, McKeever, & Schwartzman, 1984), the Texas approach to intervention as represented by two similar juvenile firesetters programs—one in Dallas (Rodrigue, 1982) and one in Houston (McKinney, 1983)—and the National Firehawk Foundation's Firehawk Children's Program (Gaynor, McLaughlin, & Hatcher, 1983). The basic features characterizing the operation of each of these programs will be described, including their goals and objectives, the types of youngsters they serve, the assessment methods they employ to evaluate youngsters, and the intervention strategies they utilize to eliminate firesetting and the accompanying pathological behavior. In addition, evaluative information regarding their effectiveness will be reviewed so that their roles can be assessed regarding the value of these intervention programs in helping to reduce recurrent firesetting and the associated psychopathology.

The Fire-Related Youth Program

Although the Rochester Fire Department's FRY Program emphasizes fire prevention, their primary goals are to identify youngsters who have engaged in unsupervised firesetting at least once, assess the risk of recidivism and need for services, and reduce future participation in nonproductive firestarting. These goals are accomplished by carrying out the program objectives of identifying, investigating, interviewing, and assessing youngsters for the purpose of establishing a disposition and referral. The goals and objectives of Rochester's FRY Program closely resemble the recommendations for developing a juvenile firesetters program outlined by the Federal Emergency Management Agency (1979, 1983).

The types of youngsters served by the FRY Program are at risk for becoming involved in future nonproductive firestarting incidents. They already have set at least one "serious" fire, many of which occurred in the home and involved structural damage. Also, many of these fires involved injuries. The majority of these youngsters do not appear to be "psychologically disturbed"; however, about one-third of them are regarded as coming from multiproblem families. The proportion of single-parent families is slightly higher than the average for the community, and the rate of unemployment for these families is extremely higher than the average. Families of these children report that firestarting materials are easily available in the home. In addition, these families report that a larger-than-average percentage of their youngsters' time is unsupervised. It is estimated that for the majority of these youngsters, removing the opportunity to start the fire should reduce the incidence, but for a smaller number of these children and families, the FRY Program also must provide a coordinated community intervention plan.

The FRY Program emphasizes an exhaustive evaluation procedure for these youngsters and their families. The evaluation is coordinated and conducted by fire investigators who have participated in "hands-on" seminars designed to teach them how to interview and evaluate children and their families. The format, structure, and content of the assessment closely follows the guidelines set up by the Federal Emergency Management Agency (1979, 1983) for assessing risk of involvement in future firesetting behavior. These guidelines suggest interviews with youngsters and their families for the purpose of gathering sufficient information about the circumstances of the firesetting, the childrens' behaviors, and the contributing environmental factors. The result of the assessment is to evaluate the nature and severity of the fire behavior and the risk of continued firestarting as well as to make a determination about the type of intervention strategy.

The primary intervention approach utilized by the FRY Program is referral. This strategy is consistent with that recommended by the Federal Emergency Management Agency (1979, 1983). The FRY Program maintains linkages with law enforcement, juvenile probation, social services, mental health, and youth services. Data from the FRY Program indicate that the majority of cases are referred to mental health agencies. The successful engagement of children and their families into their referral programs is routinely monitored and documented. The FRY Program does not attempt to provide direct intervention services; rather, it advocates strong linkages with community agencies so that the appropriate services will be available to youngsters and their families.

The rationale, organization, and operation of the Rochester Fire

Department's FRY Program is documented in a detailed report (Cole et al., 1984). In this report, statistics regarding the operation of the 1983 FRY Program are presented with information on the effectiveness of the program represented by data on the disposition of the caseload. Educational intervention was provided to 166 of the 170 youngsters, with 88 (52%) of the 170 children referred for additional services. Unfortunately, no follow-up information is reported on the outcome of either those youngsters receiving educational intervention or those referred for additional services. Because data are available regarding the firesetting histories, demographic characteristics, and existing environmental factors of this sample of 170 children and their families, it is critical that a follow-up investigation be conducted examining the outcomes of these cases. Until information is documented on the effectiveness of this community-based intervention strategy, especially with respect to the recurrence of firesetting behavior, the success of the FRY Program remains open to question.

Juvenile Firesetters Programs in Dallas and Houston, Texas

The Dallas and Houston Juvenile Firesetters Programs share common goals and objectives, work with a similar group of youngsters and their families, and employ comparable evaluation procedures, but differ somewhat on how they implement their intervention programs and how they view and treat the accompanying psychopathology. The goals of both programs are to interrupt and terminate recurrent, nonproductive firesetting and associated behavioral problems. The specific objectives of each program are to provide an effective approach to evaluating, educating, and treating youngsters and their families who are experiencing the syndrome of pathological firesetting. In addition, both programs aim at resolving underlying problems that led to the emergence of the firesetting behavior. These programs work toward elimination of nonproductive firestarting behavior and healing the underlying psychopathology.

Both the Dallas and Houston Juvenile Firesetters Programs work with youngsters who have a demonstrated history of fireplay or firesetting. These youngsters range in age from 3 to 17 and come from a variety of socioeconomic backgrounds, with the majority of them being young boys whose average age is 10 years. Arson investigators, who have identified youngsters as suspects in setting fires, are a major referral source for both these programs. The population of youngsters served by these programs want to stop their firesetting behavior but lack the necessary support and skills to interrupt what could be the beginning of a long history of antisocial behaviors.

The evaluation procedures employed by the Dallas and Houston programs are similar in both content and process. All youngsters and their families referred to these programs participate in an extensive, psychologically oriented evaluation. The evaluation methods are adopted from the guidelines published by the Federal Emergency Management Agency (1979, 1983). In addition, the evaluation process of both programs involves a built-in educational component comprised of providing printed material aimed at instructing youngsters in fire prevention and showing films designed to encourage youngsters and their families to participate in these programs. Hence, at the conclusion of this assessment, not only has there been a psychological evaluation of the presenting firesetting behavior and associated psychopathology, but youngsters have been exposed to materials designed to educate them about the rules of fire prevention and encourage them, with their family, to engage in activities that will eliminate their nonproductive firestarting and related behavioral problems.

Both the Dallas and Houston Juvenile Firesetters Programs utilize the ''graphing'' technique developed by Bumpass (1983) and described in detail in Chapter 6. The methods of implementing this procedure vary between programs. The Dallas program has trained a small, selected, prescreened group of firefighters on how to utilize the ''graphing'' method to work with children. They employ this procedure in three counseling sessions which follow their initial evaluation. The Houston program is conducted by three masters-level psychotherapists who include the ''graphing'' technique as part of their first assessment session with youngsters and their families. Although these programs adopt different methods of implementation, both utilize "graphing" as the primary method of eliminating nonproductive firestarting behavior.

The Houston program reports to offer two additional intervention strategies. First, they conduct a short-term (five-session) family therapy program designed to change the identified psychopathology accompanying the firesetting. Second, they provide youngsters with Big Brothers or Sisters who monitor their progress over a one-year period of time. This approach provides a multicomponent intervention strategy to prevent the recurrence of firesetting and its related psychopathology.

Both the Dallas and Houston programs indicate a high degree of effectiveness in eliminating firesetting behavior in youngsters. The "graphing" technique itself, employed by both programs, has been reported elsewhere as successful in eliminating nonproductive firestarting (Bumpass et al., 1983). The Dallas program showed that over a nine-month period it counseled 185 youngsters, all of whom immediately stopped firesetting, and only eight of whom had to be referred for additional types of interventions. The Houston program reports that 63 youngsters have participated in their program, and after completing one or more sessions, none of them have been involved in firesetting behavior. During this same period of time the Houston Fire Department indicates that arson losses decreased by $8,000 (McKinney, 1983). Although neither program has been operational long enough to report on the long-term success of their efforts, both report that such investigations are planned. Attempts must be made to determine not only the long-term impact of these intervention strategies on the lives of the youngsters and the families participating in such programs, but the effectiveness of these community-based interventions on reducing the incidence of juvenile-related arson, the associated property damage, and rates of injury and death.

The Firehawk Children's Program

The Firehawk Children's Program began as a research-and-development project designed to test the effectiveness of training firefighters to work with youngsters who have histories of recurrent firesetting and associated behavioral problems. The underlying assumption of the initial research-and-development project was that firefighters could work directly with youngsters to eliminate their nonproductive firestarting and remediate the accompanying behavioral problems. At the conclusion of the two-year, federally funded program development-and-evaluation phase, a nonprofit organization was created called the National Firehawk Foundation, whose mission is to distribute the Firehawk Children's Program to fire departments on a national basis as well as develop and promote educational, research, and intervention programs aimed at fire prevention and safety.

The primary goal of the Firehawk Children's Program is to provide a comprehensive intervention program to youngsters and their families designed to eliminate recurrent firesetting and the accompanying behavioral problems. This goal is accomplished by assisting fire departments to set up a volunteer-based program where firefighters work with children to accomplish the following set of objectives: (1) evaluate and identify both the youngsters' risk levels for firesetting and their dominant psychosocial needs; (2) teach basic concepts in fire safety and prevention; (3) raise feelings of self-worth by helping youngsters participate in activities appropriate to their interests and capabilities; (4) teach the "healthy" expression of feelings, especially angry feelings, and learn the appropriate expression of aggression; (5) explore recreational resources available to youngsters and their families in their communities; and (6) recommend and refer youngsters and their families for additional psychological, medical, legal, or other services on an as-needed basis. The National Firehawk Foundation provides technical assistance to operationalize these goals and objectives in fire departments around the country.

The Firehawk Children's Program works with a similar population of youngsters and families as the previously described participants of these early identification and intervention programs. Arson investigators and parents typically refer youngsters to the Firehawk Children's Program. A preliminary survey sampling 10 youngsters in four geographical sites operating Firehawk Children's Programs showed that 85% were males with a modal age of 10. The majority of the boys were white, reflecting the proportional ethnic distribution of the four communities. One of the four communities was predominantly black, and these ethnic statistics were represented in the population served by their Firehawk Children's Program. The participants in each of the four programs were mostly youngsters coming from single-parent families (76%) who had expressed their interest in fire by engaging in at least one unsupervised incident of fireplay, which may or may not have resulted in an actual fire. An analysis of the accompanying psychosocial behavioral correlates of these youngsters as reported by their parents indicates a significant occurrence of the expression of aggression or anger, behavior problems in school and family discord, with one or more parents absent for extended periods of time. These characteristics of the youngsters and families participating in Firehawk Children's Program are consistent with the description of population served by other early-identification-and-intervention programs indicating that behavioral and family pathology accompanies the occurrence of firesetting behavior.

The Firehawk Children's Program utilizes a two-phase evaluation system for youngsters and their families. When a referral is made to the program, an initial screening is conducted by a trained firefighter. This preliminary assessment follows the guidelines specified by the Federal Emergency Management Agency (1979, 1983) and includes an analysis of firesetting risk levels as well as a description of the individual and social characteristics of youngsters and their families. In addition, during the initial evaluation, educational materials describing important fire prevention rules are distributed, and all youngsters are invited to return to the fire department for a workshop focused on fire safety. The result of the preliminary assessment is the determination of firesetting risk levels and the disposition for intervention. For all cases this analysis is reviewed by a mental health professional. For those youngsters exhibiting a "definite" risk for firesetting (about 40% of the cases), the mental health professional conducts a second evaluation emphasizing underlying psychosocial factors contributing to the cause of the firesetting behavior. The result of this second screening is the identification of the determinants of the nonproductive firestarting behavior and the design of a specific, long-term intervention strategy to remediate the observed psychosocial problems.

In addition to the educational intervention provided to all youngsters evaluated by the Firehawk Children's Program, once firesetting risk level has been determined, there are a number of different intervention options. If there is relatively little risk for future firesetting behavior, an educational seminar is offered which teaches fire safety and prevention and which is conducted at the local fire station. If there is a definite risk for firestarting, then volunteer firefighters are assigned to work on a long-term basis with youngsters and families. The role of the firefighters is multidimensional, but firefighters are trained in how to establish and maintain close, effective, working relationships with youngsters. Firefighters learn how to be role models for these youngsters while at the same time being someone with whom these youngsters can talk and spend time. The objectives of this partnership between firefighters and youngsters is to eliminate firesetting and redirect this nonproductive expression of aggression toward more positive outlets such as sporting activities and other positive recreational endeavors. For definite as well as extreme firesetting-risk youngsters, the Firehawk Children's Program builds a solid network of community resources so that the variety of identified needs these youngsters and their families express can be addressed by effective and appropriate referral. Hence, the Firehawk Children's Program offers comprehensive intervention options which include—subsequent to a thorough evaluation—education, working partnerships with firefighters, and referral to supporting community agencies.

During the two-year research-and-development phase of the Firehawk Children's Program, a cost-effectiveness study was conducted to determine the relative costs to fire departments and communities in providing an early identification and intervention program and the effectiveness of such a program in reducing nonproductive fireplay and firesetting. One hundred and eighty youngsters and families were screened during the study period and classified according to the Federal Emergency Management Agency guidelines (1979, 1983) as little risk (66%), definite risk (31%), and extreme risk (3%) for future firesetting involvement. The youngsters and their families were monitored on cost and effectiveness dimensions as they entered the program and during the two-year study period. Effectiveness was defined as the impact of the Firehawk Children's Program in reducing the incidence of firesetting and improving the quality of life for definite-risk youngsters as indexed by expected behavior changes, including an increase in emotional expression, an increase in impulse control, and an increase in involvement in recreational activities. Costs were summarized in three categories—activities for youngsters, training for firefighters and staff, and educational and other materials—and were calculated including additional ''startup'' costs such as program development and research and evaluation, many of which are one-time expenditures and therefore slightly inflate the figures. Based on reported incidents of recurrent firesetting and a total program cost for one year of $8,000 for servicing a 650 million person metropolitan community, a cost-effectiveness ratio was derived for the Firehawk Children's Program. The program costs during the research-and-development phase compared to the number of mean fire incidents potentially prevented as a result of youngsters participating in the program reflect a cost-effectiveness ratio of $41.94 per fire prevented. It is known that the youngsters entering the program averaged 310 firestarts, the majority of which resulted in no property loss, damage, or injury. However, one school fire was set and resulted in damages totaling $300,000. Hence, the estimated cost of $41.94 per fire prevented may be well worth the investment in the Firehawk Children's Program to prevent just one large, costly fire.

The National Firehawk Foundation reports that they have planned follow-up studies utilizing the multiple sites that have set up and maintained the Firehawk Children's Program since the initial phase of research and development. However, they, like the other early-identification-and-intervention community programs, have yet to offer information on the effectiveness of their efforts in reducing the incidence of juvenile-related arson. In addition, although preliminary evidence indicates the successful elimination of firesetting coupled with an improved ''quality of life'' for youngsters participating in the Firehawk Children's Program, there are no long-term data to support sustained near-zero recidivism and significant changes in the associated behavioral patterns. The Firehawk Children's Program needs to pay further attention to specifying its long-term impact on reducing the incidence of juvenile arson and remediating the accompanying psychosocial determinants.

The Application of Early-Identification-and-Intervention Programs

Table 7.2 summarizes the essential features characterizing community-based programs designed to provide early-identification-and-intervention services to youngsters and families presenting with recurrent firesetting. These features include for each program their goals and objectives, the types of youngsters and families they serve, the evaluation methods they utilize to assess youngsters, the intervention strategies they employ, and the effectiveness of these programs in eliminating nonproductive firestarting and the associated psychopathology. All of these early-identification-and-intervention programs operate in conjunction with local fire departments primarily because they are the predominant community agency to have first contact with youngsters suspected of being involved in firesetting behavior. Therefore, fire departments, networking with other local services and agencies, are responsible for the organization and operation of these community-based intervention programs.

All of these early-identification-and-intervention programs share similar goals and objectives, service a population of youngsters and families characterized by a common set of demographic factors, and apply uniform guidelines for evaluation. The major differences between these programs are the intervention strategies they utilize to eliminate recurrent firesetting and the accompanying behavioral problems. The interventions range from providing evaluation and referral services, as represented by Rochester's FRY Program, to offering a full-range of direct services such as psychotherapy and partnerships, as reflected by the Dallas, Houston, and Firehawk Children's Programs. The selection of the appropriate community-based program depends on optimizing the match between the psychosocial needs of the youngsters and their families and the specific methods of intervention employed by these community-based programs. For example, if a psychological evaluation reveals

Table 7.2

The Major Factors Characterizing Early Identification and Intervention Programs

Program	Goals/Objectives	Target Population	Evaluation Methods	Intervention Strategies	Effectiveness
Fire-Related Youth Program	To identify youngsters who have set at least one unsupervised fire, assess their risk for recidivism and need for services, and prevent future firestarts	"Normal" youngsters coming from "multiple-problem" families experiencing a higher-than-average unemployment rate	An evaluation conducted by trained fire investigators to assess the firesetting risk level of youngsters following the guidelines of the Federal Emergency Management Agency	Evaluation, distribution of educational materials, and referral for direct services	Disposition of caseload indicates that 97% of the youngsters received educational intervention and 52% of the youngsters were referred for additional services
Dallas and Houston Juvenile Firesetters Programs	To interrupt and terminate recurrent firesetting and associated behavioral problems by evaluating, educating, and treating youngsters and their families	Young boys between the ages of 3 and 17 who come from a variety of socioeconomic backgrounds and who express a desire to stop firesetting	An evaluation conducted according to the guidelines of the Federal Emergency Management Agency; includes educational materials designed to "encourage" compliance with fire safety	The application of the "graphing" technique; the Houston Program offers short-term family psychotherapy and Big Brothers or Sisters	No recidivism reported by both programs; the Dallas Program indicates that only 23% of their youngsters had to be referred for additional services; the Houston Program reports an $8,000 decrease in arson losses
The Firehawk Children's Program	To provide a comprehensive intervention program to eliminate recurrent firesetting and accompanying behavior problems by providing evaluation, education, intervention focused on a partnership between youngsters and firefighters and referral services	Young boys with a modal age of 10, coming from single-parent families, who have engaged in at least one unsupervised firestart and who show significant expressions of inappropriate aggression and anger	A two-phase evaluation procedure where the first phase follows the guidelines of the Federal Emergency Management Agency and the second phase consists of a psychosocial assessment conducted by a mental health professional	Fire safety and prevention education seminars, partnerships between youngsters and trained firefighters, and referral services	Research-and-development phase indicated no recidivism, improved "quality-of-life" scores, and a cost-effectiveness ratio of $41.94 per fire prevented

that nonproductive firesetting began shortly after the occurrence of several stressful family events, including separation and divorce, then the selection of an appropriate community-based program must take into account the amount of attention that will be paid to working with the family system. The community-based programs that offer direct family psychotherapy services (the Houston Program) or a long-term partnership between younsters and firefighters (the Firehawk Children's Program) may be the most effective intervention strategy when family stress is one of the precipitating factors in the onset of nonproductive firestarting.

The selection of effective community-based programs must be based not only on the type of intervention strategy utilized, but on evidence indicating the success of their efforts in eliminating recurrent firesetting and the accompanying psychopathology. Preliminary data indicate that these early-identification-and-intervention programs are successful in stopping nonproductive firestarting. There is less information available regarding their effectiveness in remediating the psychopathology associated with firesetting. Both the ''graphing'' technique employed by the Dallas and Houston programs and the partnership method utilized by the Firehawk Children's Program have fared well in follow-up studies. However, because these investigations do not include comparison groups, it remains unclear whether reported positive change in ''quality-of-life'' variables such as emotional expression, impulse control, and participation in recreational activities is a real effect of these programs. In addition, there is no evidence on the long-term impact of these community-based programs in reducing the incidence of juvenile-related arson. Until these issues are clarified, the relative success of these early-identification-and-intervention community-based programs in eliminating firesetting behavior and remediating the accompanying psychopathology remains open to future research and evaluation.

COMMUNITY INTERVENTION AND PSYCHOTHERAPY

Regardless of the specific strategy, the primary goal of any community intervention or psychotherapy effort is to reduce incidents of unsupervised fireplay, nonproductive firestarting, and juvenile-related arson. Secondary goals include remediating accompanying psychopathology and teaching youngsters to demonstrate fire-safe behavior. Given these significant goals of survival and psychological well-being, a model is suggested that outlines a developmental plan of intervention for youngsters. This proactive model is designed for the majority of youngsters, utilizes a developmental perspective, and includes the basic components of prevention and, when indicated, community-based intervention and psychotherapy. The application of such a model raises important research issues regarding intervention strategies aimed at youthful fire behavior.

A Proactive Model

The major reason for proposing a proactive model is to synthesize what is currently "known" about how to eliminate the occurrence of pathological firesetting and remediate the accompanying behavioral correlates. The predictive model presented in Chapter 3 organizes the existing knowledge base for the purpose of explaining youthful fire behavior and anticipating the emergence of pathological firesetting. This proactive model seeks to identify guidelines on how to increase the probability of youngsters demonstrating fire-safe behavior and how to decrease the probability of a major fire occurring as the result of unsupervised fireplay or nonproductive firestarting. Hence, the predictive model describes the development of youthful fire behavior and anticipates the emergence of pathological firesetting, and the proactive model identifies the prevention methods to teach fire-safe behaviors and outlines intervention strategies to eliminate nonproductive firestarting and the accompanying psychosocial determinants.

Like the predictive model, the proactive model is based on the assumption that fire interest naturally occurs in most youngsters as early as age three. Therefore, this model indicates that prevention programs must be introduced coincidentally with the initial emergence of fire interest. However, prevention efforts must begin at the preschool level and be carried on consistently and with age-appropriate instruction throughout elementary-school age and into early adolescence. It is expected that continued exposure to increasingly sophisticated fire safety rules and behaviors will deter the majority of youngsters from participating in unsupervised fireplay and nonproductive firestarting.

There is a proportion of youngsters who will participate in prevention programs, but who also will experience a set of psychosocial conditions that will influence them to become involved in pathological firesetting behavior. For most of these youngsters, the probability is high that their nonproductive firestarting can be interrupted and terminated and the accompanying psychopathology can be remediated if these problems are identified early and they receive the appropriate intervention. Early detection, a complete evaluation, and the selection of the optimal intervention strategy are the essential elements for assuring these youngsters and families the best chance for a successful recovery.

The current intervention options designed to eliminate firesetting and remediate the accompanying psychopathology must be given careful consideration. The question is not whether a community-based program or psychotherapy is more appropriate, but what combination of intervention strategies will yield the best result. Some community-based programs are designed to be used in conjunction with psychotherapy, such as Rochester's FRY Program, some offer the direct services of specific types of psychotherapy, such as the Dallas and Houston programs, and others may offer alternatives or adjunct methods of treatment, such as the Firehawk Children's Program. Participation in psychotherapy does not preclude participation in a community-based program. The selection of an effective intervention strategy is a clinical decision-making process based on a thorough analysis of evaluation data and a matching of the psychosocial profiles of youngsters and families to the specific objectives and methods of community-based programs and/or psychotherapies. The desired outcome from this optimal match is elimination of recurrent firesetting behavior and the accompanying psychopathology.

Figure 7.1 summarizes the major components of this proactive model. The three primary features of this model are prevention, detection, and intervention. This model indicates that for the majority of youngsters, prevention efforts, beginning with the emergence of fire interest and continuing through elementary-school age and early adolescence, should successfully deter participation in

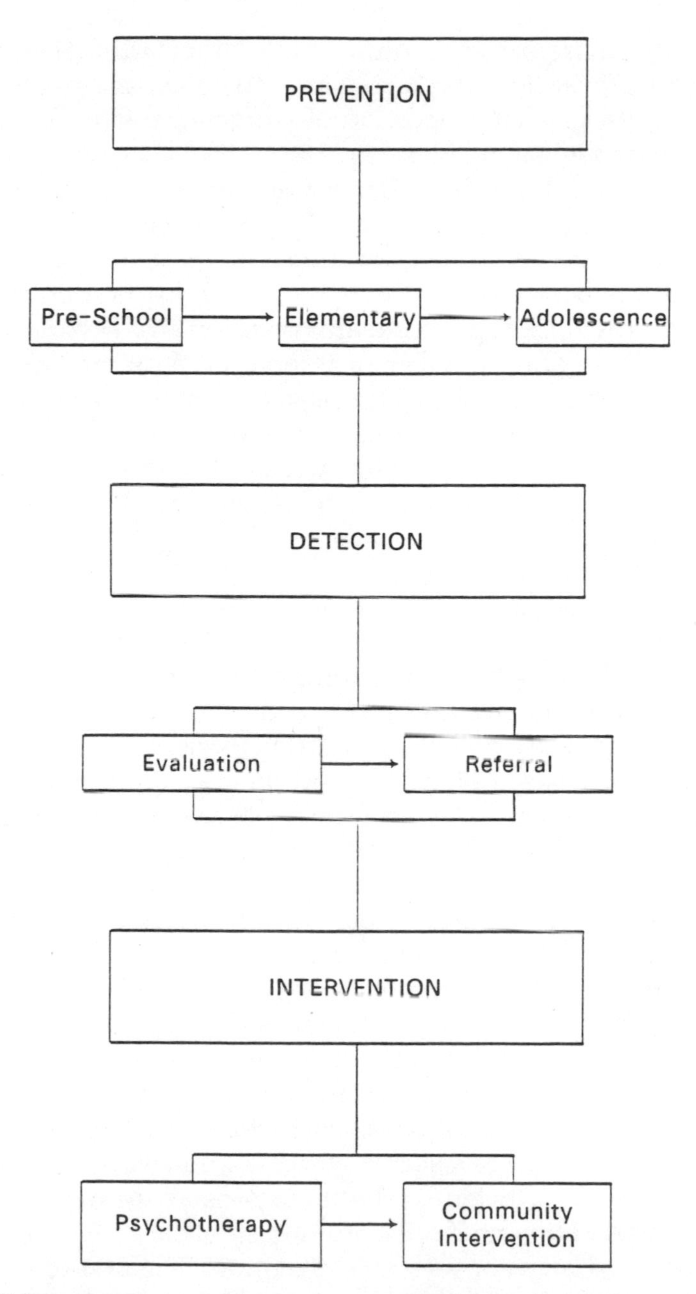

Figure 7.1. A proactive model to reduce the incidence of unsupervised fireplay, nonproductive firestarting and juvenile-related arson.

unsupervised fireplay and nonproductive firestarting. Despite exposure to prevention programs, and given certain psychosocial determinants, a small proportion of youngsters will express their fire interest by engaging in unsupervised fireplay or recurrent firesetting behavior. Early detection coupled with a thorough psychological evaluation is necessary to identify the severity of the firesetting and the accompanying psychopathology. Once the firesetting behavior has been assessed and the psychosocial determinants identified, then the appropriate intervention strategy must be set in place. There are a number of intervention options, including community-based programs and psychotherapies. The successful selection of an intervention strategy depends on optimizing the match between the psychosocial needs of youngsters and their families and the objectives and methods employed by intervention programs.

Intervention Research

The most dramatic deficit in attempts to deter unsupervised fireplay and nonproductive firestarting is the lack of research focused on evaluating the effectiveness of current intervention efforts—both the community-based programs and the psychotherapies. Major work must be undertaken to compare the relative effectiveness of the various community-based intervention programs as well as the relative effectiveness of the various psychotherapies. Not only must effectiveness issues be examined, but the relative costs of these intervention strategies must be evaluated. Only with the application of rigorous clinical research standards can excellence be clearly identified with respect to successful intervention programs designed to eliminate firesetting and the accompanying psychopathology.

Although there is the potential for a great many clinical research comparisons to be made, it is within the framework of this analysis to suggest what type of research effort would best represent "the edge of the field." Obviously, the characteristics distinguishing the various intervention strategies are not clear-cut, but at least two fundamental dimensions deserve attention. The first of these dimensions which is relevant for a clinical research investigation is the comparison between community-based intervention strategies and psychotherapies to treat firesetting behavior and the accompanying psychopathology. Clearly, within each of these approaches, there exists a diverse application with respect to philosophies and methods of intervention. However, the underlying assumptions and diverse programmatic features characterizing these two strategies warrant the scrutiny of a controlled research investigation. Community-based programs offer evaluation, education, and, for a specific population of identified youngsters, an intervention delivered by trained firefighter personnel focused on eliminating firesetting behavior and in some instances the accompanying psychopathology. Psychotherapies, delivered by mental health professionals, also offer evaluation coupled with a particular philosophy and technique to stop nonproductive firestarting and remediate the associated behavioral problems. The short- and long-term effectiveness as well as the costs of these two somewhat diverse intervention strategies must be compared utilizing adequate clinical research standards such as appropriate comparison groups, random assignment, and accurate outcome measurement.

The second dimension distinguishing these intervention strategies is the relative attention they pay to the psychopathology that typically accompanies firesetting behavior. It is significant to note that all of the intervention strategies, community-based programs and psychotherapies, are reportedly successful in eliminating firesetting behavior. One of the major differences is the assumption they make about the underlying causes of the firesetting behavior and consequently the types of psychosocial determinants they recognize as significant. The relative importance placed on the psychosocial determinants of firesetting behavior is directly related to their methods of intervention. For example, the ''graphing'' technique employed in both psychotherapies and community-based programs assumes that the emotional state and critical events immediately preceding the firesetting incident are the most significant causative factors and hence become the focus of intervention. Other approaches assume that the occurrence of firesetting behavior in youngsters reflects the underlying pathology in their social relationships, in particular those relationships within the family. Therefore, family psychotherapy becomes the intervention of choice. The key research issue is identification of the types of intervention approaches that are effective in remediating the relevant psychosocial determinants underlying the firesetting behavior. In addition, there may be particular types of youngsters and families who may respond better to specific intervention strategies. A clinical research study focused on different intervention approaches not only must compare their relative effectiveness in eliminating firesetting, but it must assess their relative emphasis on treating the identified psychopathology that accompanies the behavior. Measurement of the specific effects of comparative interventions will yield significant information on the psychosocial determinants of firesetting.

SUMMARY

The two major types of community-based efforts designed to reduce unsupervised fireplay and nonproductive firestarting—prevention and early-identification-and-intervention programs—have been implemented utilizing a variety of philosophies and techniques. Prevention programs operate under the general assumption that fire interest is a naturally occurring curiosity and therefore the rules and behaviors of fire safety must be taught to all youngsters. There are a number of prevention programs in school settings, ranging from those aimed at preschool-age children to those instructing adolescents how to be role models and teachers of fire safety and prevention to fourth- and fifth-grade youngsters. Although research evidence is sketchy, it is suggested that if youngsters learn age-appropriate fire-safe behaviors through prevention programs, the probability is greatly reduced that they will engage in unsupervised fireplay and nonproductive firestarting. Despite prevention efforts and because of the occurrence of a specific set of psychosocial circumstances, a proportion of youngsters will become involved in recurrent, pathological firesetting behavior. If early identification and intervention procedures can be initiated for these youngsters and their families, then the probability is high that their nonproductive firestarting will be eliminated and the accompanying psychopathology will be remediated. There are a variety of community-based intervention strategies which operate mostly in fire departments and provide a spectrum of services including evaluation, education, psychotherapy, partnership, and referral. A successful treatment outcome for pathological firesetting depends on optimizing the match between the psychosocial profile of youngsters and their families and the appropriate intervention—whether it be a community-based program, psychotherapy, or some combination of strategies. In general, a number of proactive steps can be implemented to reduce the likelihood of youngsters becoming involved in firestarting and juvenile-related arson. However, the dramatic deficit in clinical research focused on evaluating the relative effectiveness and cost of community-based programs and psychotherapies makes it impossible to clearly identify excellence with respect to the successful treatment of pathological firesetting.

8

Working with Youthful Firesetters—Multiple Roles and Responsibilities

Effective detection and intervention strategies aimed at eliminating recurrent firesetting and the accompanying psychopathology are enhanced by mental health professionals who are aware of the multiple roles and responsibilities they can assume when delivering these services. There are a number of traditional as well as innovative roles for mental health professionals, including those of clinician, consultant, expert witness, and advocate. Each of these roles carries with it specific functions which represent various types of professional activities. In addition to these multiple roles, there are a variety of responsibilities which are specific to working with youthful firesetters and their families and which are important concerns for mental health professionals. Among these responsibilities are issues of confidentiality, including ''reasonable'' trust in verbal communication, the privacy of written records and identification, and disclosure through publications and the media; and professional liability, including insurance and legal matters. In addition to the successful application of these professional roles and responsibilities, mental health care providers must work toward a common set of goals which will eliminate nonproductive firestarting and make a significant contribution to understanding the psychology of child firesetting.

ROLES—TRADITIONAL AND INNOVATIVE

Mental health care professionals can assume a number of different roles in working with youthful firesetters and their families. The four predominant roles in which professionals are likely to find themselves are clinician, consultant, expert witness, and advocate. Specific functions are attributable to each of these roles, which identify particular professional activities. For example, the clinician's role is typically characterized by the functions of evaluation and treatment, whereas the consultant's role may consist of program development, education, or training activities. These specific functions identify a variety of ways in which mental health care professionals can contribute to the study and treatment of youthful firesetting.

The four predominant roles and their functions can be viewed as both traditional and innovative. The traditional roles are identified by a set of familiar activities in which mental health care professionals typically participate. For example, the role of clinician, as characterized by the functions of providing psychological evaluation and treatment to youthful firesetters and their families, is a well-accepted standard of professional practice. However, the consultant's role, as defined by such activities as working with fire departments and community agencies to develop screening and counseling programs for youthful firesetters and their families, is a rather unique function for mental health care professionals. There is a significant value attached to those traditional roles which maintain the acceptable standards of professional practice as well as those innovative roles which break ground in defining a new set of professional activities.

The four roles—clinician, consultant, expert witness, and advocate—and their specific functions will be described in terms of the particular types of activities they offer for mental health care professionals. The traditional as well as innovative functions will be considered with respect to their value in contributing to the understanding of the psychology of child firesetting.

Clinician

The two major functions characterizing the role of clinician are evaluation and treatment. The mental health care professional can contribute to direct service delivery by providing psychological evaluation and treatment to youthful firesetters and their families. These services represent the unique activity of mental health care professionals in offering effective intervention strategies designed to eliminate recurrent firesetting and the accompanying psychopathology.

Previous chapters have described the clinical characteristics of pathological fire behavior, outlined methods for interviewing and evaluating youthful firesetters and their families, and detailed the various outpatient and inpatient options for psychotherapy. This information is based on a current assessment of the ''state of the art'' in providing effective evaluation and treatment services to youthful firesetters and their families. Mental health care professionals assuming the role of clinician and performing the functions of evaluation and treatment can select from this menu of current detection and intervention methods.

In addition to these acceptable standards of clinical practice, mental health care professionals dedicated to eliminating firesetting behavior and the accompanying psychopathology must search for new methods of clinical treatment. Given the current ''state of the art'' of clinical practice, there is room for significant advances in developing new techniques for stopping nonproductive firestarting and remediating the associated psychosocial correlates. In particular, clinicians must pay closer attention to discovering how to adjust those individual characteristics, social circumstances, and environmental conditions which are so closely related to the emergence of pathological firesetting. Hence, while mental health care professionals must continue to participate in the delivery of traditional clinical services, they also must engage in the design of innovative methods of clinical evaluation and treatment.

Consultant

There are a variety of potential functions to be carried out by mental health care professionals who want to provide consultation services. The two most frequently requested activities are program development and providing educational and training services. These functions generally are carried out by mental health care professionals within the context of fire departments and community agencies.

Because of the recent interest on the part of fire departments and community agencies in developing early identification and intervention programs as described in Chapter 7, there is a demand for mental health care professionals to provide technical assistance to design, develop, and implement these services. The successful delivery of effective early-identification-and-intervention programs by fire departments and community agencies requires the expertise provided by mental health care professionals. In addition, it may be to the benefit of those clinicians treating youthful firesetters and their families to engage the interest of their local fire departments and related service agencies to provide additional resources such as community-based intervention programs to assist in combating the problem of youthful firesetting.

Along with program development within the community, mental health care consultation services are necessary to educate and train those who are interested in helping youthful firesetters and their families. Among the two groups requiring educational and training programs are other mental health care providers, whether they are in private practice or working in community agencies, and fire department personnel, who primarily are responsible for operating community-based intervention programs. Depending on the various activities in which these groups will engage, they will need education and training programs focused on how to screen and evaluate youngsters presenting with nonproductive firestarting behavior, how to implement specific intervention strategies, and how and where to refer these children and their families for additional services. The delivery of relevant and effective education and training programs by mental health care professionals will help build a substantial system of qualified services for youthful firesetters and their families.

The role of providing the consultation functions of program development and educational and training services to other mental health care providers, fire departments, and related community agencies is a comparatively new and unique set of professional activities. The opportunities are significant for developing important new systems of service delivery to help youthful firesetters and their families. In addition to the benefits of building effective detection and intervention programs, a cooperative relationship is established between mental health and the fire service.

Expert Witness

Because mental health care professionals are achieving a certain level of competence in identifying and treating youthful firesetters, and because youthful firestarting is a potential criminal behavior, there has been a growing demand for expert testimony in juvenile-related arson cases. Whether testifying on behalf of the prosecution or the defense, there is a certain set of specialized skills mental health care professionals can bring to the courtroom. These specific functions comprise the role of expert witness that can be assumed by mental health care professionals.

The specific functions of mental health expert witnesses are their ability to apply mental health data, professional findings, and clinical inferences to identify mental responsibility as it relates to the criminal behavior of arson. The legal definition of arson, as malicious intent to destroy by fire, and the six elements comprising mental responsibility are presented in Chapter 2. If the physical evidence leaves little question as to the guilt of youngsters committing arson, then the issue remains as to whether juveniles, especially those 14 years and younger, are mentally responsible for their actions. The issues involved in making such a determination also are reviewed in Chapter 2. The role of mental health care professionals is to provide a comprehensive psychosocial evaluation, including a thorough history of fire behavior, and through an analysis of these data arrive at a recommendation as to mental responsibility.

The critical task of court evaluations is integration of psychosocial and firesetting histories to determine whether the specific act of arson under consideration was executed with mental responsibility. Chapter 5 outlines interview methods and the relevant clinical data that must be collected during such an assessment. However, the fundamental difference distinguishing routine psychosocial evaluations from those which will be reviewed by the court is the interfacing of the data obtained from the psychosocial and firesetting histories with the six elements defining mental responsibility. Each clinical decision must be made on a case-by-case basis, matching the criteria of mental responsibility with what is known about the youngster's consciousness, intention, understanding, and control of behavior. This type of clinical decision making, especially when the concern is youngsters under 14 years of age, is perhaps one of the most challenging functions of the expert witness.

The results of the evaluation and analysis of psychosocial and firesetting data as they relate to identifying the mental responsibility of youngsters accused of participating in the crime of arson generally are presented in writing to the court. Once these reports are entered as evidence, the responsible mental health care professional may be asked to testify, although these reports often are taken at their face value when juveniles are the concern. However, mental health care professionals must be prepared to present their findings verbally in court. Therefore, it is important to be able to write as well as articulate an accurate analysis of mental responsibility as it relates to arson crime.

The role of expert witness and the accompanying functions of being able to provide written and verbal evaluations of youngsters' mental responsibility in the crime of arson are new professional activities for mental health care providers. The stresses unique to courtroom procedures, especially where youngsters' futures are being decided, may influence whether mental health care professionals choose to become involved in these types of activities. There are some challenges, such as participating in the determination of youthful criminal behavior, as well as some benefits, such as making a contribution toward rehabilitation, which must be considered when assuming the role and associated functions of the expert witness.

Advocate

Of all of the four roles—clinician, consultant, expert witness, and advocate—the role of advocate is perhaps the newest and, as such, the least well defined of these professional responsibilities. By definition, the role of advocate is one in which professional activities are focused on supporting programs that help to understand and rehabilitate youngsters involved in pathological firesetting. The functions characterizing the role of advocate are those activities aimed at initiating changes in the current social policy which affects the treatment of firesetting youngsters. The two major activities that potentially can result in changes in social policy are direct participation in political and governmental processes and execution of relevant research projects. These professional activities represent just two methods of influencing the current social policy as it relates to the treatment and rehabilitation of youthful firesetters.

Participation as an advocate for youthful firesetters in political and government processes is a unique professional activity. There are multiple avenues of concern to be supported at the local, state, and federal levels on behalf of youthful firesetters. There are some recent examples of how effective the advocacy role can be at the state and federal levels in terms of making modest inroads into the legislative process with respect to recognizing the importance of juvenile firesetting. At the state level, the 1985 California legislature passed a bill, supported by Senator William Campbell (R, Orange County) and the National Firehawk Foundation, to establish a statewide Child Arson Task Force to study the problem of juvenile firesetting (Chapter 1529, California State Code, 1986). The introduction of a similar bill is being initiated in several other states including New Jersey and Florida. At the federal level, U.S. Senator Arlen Specter (R, Pennsylvania), also with the support of the National Firehawk Foundation, sponsored a hearing before the Subcommittee on Juvenile Justice of the Committee on the Judiciary of the United States Senate on the Problem of Arsons Committed by Juveniles (Committee on the Judiciary, 1985). Assisting in the making of new legislation and organizing and testifying at hearings represent important advocacy functions in which mental health care professionals can participate to influence the current social policy related to juvenile firesetting.

The second activity that potentially can result in changes in social policy is that of research. Although the role of advocate and the function of research traditionally are not viewed as similar professional activities, the topic and focus of research and how the results are utilized can be relevant to changes in social policy relating to juvenile firesetting. For example, research aimed at determining the most effective methods for rehabilitating youthful firesetters may influence the current criminal-legal policy of counsel and release as described in Chapter 2. Therefore, conducting or supporting applied research projects and participating in activities that utilize results to effect changes in current social policy related to the treatment of juvenile firesetters are significant advocacy functions for mental health care professionals.

The specific functions of the role of advocate—direct participation in political and governmental processes and execution and support of applied research projects—represent activities that champion the cause of child advocacy with respect to how current social policy affects the treatment of juvenile firesetters. Therefore, many mental health care professionals may not want to "take sides" for or against issues relating to child firesetting and the social policy decision-making process. Mental health care professionals individually must decide to what extent, if any, they are willing to participate as advocates for the humane treatment of youngsters involved in the potential criminal behavior of firesetting.

Roles and Functions

Table 8.1 summarizes the roles and associated functions in which mental health care professionals can participate to understand, rehabilitate, and prevent the problem of child firesetting. This list of roles and functions is not meant to be exhaustive, but rather it

Table 8.1
Roles and Functions for Mental Health Care Professionals

Role	Functions
I. Clinician	A. Evaluation
	B. Treatment
II. Consultant	A. Program development
	B. Education and training
III. Expert witness	A. Court evaluation
IV. Advocate	A. Political and government activities
	B. Research

should be regarded as a starting point from which mental health care providers can explore their options for professional involvement. The roles of clinician, consultant, expert witness, and advocate represent both traditional and innovative functions and activities for mental health care professionals. From delivery of direct services, such as psychological evaluation and treatment, to participation in government processes, such as sponsoring new legislation, mental health care professionals can make significant contributions to reducing the problem of child firesetting and providing effective rehabilitation services for youngsters and their families.

PROFESSIONAL RESPONSIBILITIES

Along with the roles and functions available to mental health care professionals come significant professional responsibilities which must be taken into consideration when working with youthful firesetters and their families. The two major classes of professional responsibility that deserve special attention are confidentiality and liability. Because of the interface between the mental health care and juvenile justice systems regarding the evaluation and treatment of juvenile firesetters, issues of confidentiality become important for mental health care professionals as well as for youthful firesetters and their families. In addition, because several indicators suggest that these youngsters are at risk for participating in future firestarts, a potentially criminal as well as destructive behavior, questions regarding the extent and degree to which mental health care professionals are liable must be addressed. Confidentiality and liability issues are significant as they specifically relate to professional activities involving child firesetters and their families.

The three essential topics concerning confidentiality are the notion of reasonable trust regarding verbal communication, the privacy of written records, and identification and disclosure considerations regarding publications and the media. Each of these topics involves a close examination of the relationship between mental health care professionals and their clients. In general, clients are youngsters and their families. However, there are certain circumstances where mental health care professionals are under contract with private or public community agencies and therefore are responsible, in part, to adhere to issues of confidentiality as they relate to these community agencies. The rules of confidentiality primarily are set up for the protection and benefit of clients.

The two critical concerns regarding professional liability are insurance and legal matters. It is necessary to understand the potential professional risks that might be incurred while working with juvenile firesetters and their families. In addition, mental health care providers must learn how to protect themselves from incurring these risks. An understanding of the insurance and legal issues that might arise and how to work effectively within their parameters will ensure a successful and rewarding professional career working within the field of child firesetting.

The questions of confidentiality and professional liability emerge directly from the context of the four predominant roles—clinician, consultant, expert witness, and advocate—which mental health care professionals are likely to assume while working with youthful firesetters and their families. Hence, the type of role and associated functions determine, to some extent, the kind of professional responsibilities that are likely to be important. For example, if mental health care professionals provide psychological evaluation and psychotherapy services to firesetting youngsters and their families, the issue of what defines a reasonable trust of verbal communication between clinicians, youngsters, and family members becomes relevant to all those participating in the therapeutic relationship. The successful definition of reasonable trust will help solidify the working relationship between clinicians and clients. The effective execution of these four predominant roles will depend on an understanding of the inherent professional responsibilities.

Confidentiality

The issues of reasonable trust of verbal communication, the privacy of written records, and identification and disclosure considerations regarding publications and the media are the three most frequently occurring topics of confidentiality when working with youthful firesetters and their families. There are rules of confidentiality which guide the actions of mental health care professionals. These guidelines protect the nature of the professional-client relationship and, within legal parameters, allow for an effective relationship between firesetting youngsters, their families, and mental health care professionals.

Establishing reasonable trust in verbal communication is critical in all four professional roles. However, it is particularly relevant in those roles—clinician and expert witness—where mental health care professionals provide direct services, such as psychological evaluation and psychotherapy, to firesetting youngsters and their families. During an evaluation or treatment session youngsters may want to confide in clinicians and tell them things they do not want to tell their parents. In addition, parents may put pressure on clinicians to tell them all about what is happening with their youngsters, or they may share with clinicians information they do not want their youngsters to know. It is important to clarify the rules of reasonable trust as they relate to confidentiality early in the therapeutic relationship.

One goal of an effective relationship is to develop a reasonable trust between clinicians, firesetting youngsters, and their families. Therefore, in the early stages of a working relationship, the idea should be presented to both youngsters and family members that there may be thoughts and feelings shared between clinicians and youngsters or clinicians and other family members that are shared in confidence. It should be understood that participants have a right to private thoughts and feelings and that confidential communications will not be disclosed unless it is in the best interest of youngsters or their families. In addition, it should be understood that disclosure of a confidence will not occur without that person's knowledge. That is, if information is shared in confidence, the trust will be kept unless there is an important reason, such as a threat to firestart for the malicious intention to destroy or burn, to break the trust. Before such a trust is broken, the youngsters or family members whose confidence is being broken should be so informed and reasons why should be made explicit. Absolute confidence in any relationship is an unrealistic goal, but building a reasonable trust is laying the necessary groundwork for an effective therapeutic relationship.

Once a reasonable trust has developed, the next step is to protect the confidence. Firesetting is often an embarrassing and painful event in the lives of youngsters and their families. The majority of citizens living in the community, including relatives and close friends of families, do not easily understand or accept firesetting behavior. Therefore, there are circumstances in which youngsters and their families may want their privacy protected. For example, if during the process of a psychosocial evaluation it is necessary to contact the school to obtain accurate information on behavior or academic performance, it is necessary to first speak with parents about this and obtain written permission from them regarding the release of the information. Issues of whether school authorities know about the firesetting behavior and whether they "should" or "need" to know all must be discussed with parents. Although open communication channels are helpful to everyone helping to work out a problem, especially where youngsters are concerned, the potential risk is that disclosure of certain types of information, such as a history of firestarting, may negatively "label" youngsters and deprive them of future learning or work opportunities that are far removed from their current difficulties. Therefore, in each instance in which there is potential to disclose information regarding the firesetting behavior of youngsters, parents need to be informed, and issues of privacy and confidentiality must be carefully considered on behalf of youngsters and their families.

The rules of reasonable trust of verbal communication are different when mental health care professionals are conducting court evaluations for firesetting youngsters and their families. These rules are dictated by who will be the recipient of the expert witness's evaluation. What information is disclosed and how it is presented depends, in part, on how it will be applied. All of the participants, the mental health care professionals, the firesetting youngsters, their families, and the defense or prosecution must understand, beforehand, the rules of confidentiality as they relate to court evaluations. Once these rules are explicit, there is less likely to be a misunderstanding or misinterpretation of the information disclosed in the psychosocial evaluation.

Along with the importance of establishing reasonable trust in verbal communication, written records also become a focus for the rules of confidentiality. In this instance all four roles—clinician, consultant, expert witness, and advocate—are likely to be affected because records generally are kept in relation to the types of activities performed. However, the records kept on behalf of firesetting youngsters are likely to be the most sensitive because they contain information on a variety of their behaviors, some of which may have criminal implications. Therefore, it is important that specific safeguards be set up to ensure the privacy of records involving firesetting youngsters.

Records detailing the psychosocial and firesetting histories of youngsters are likely to be found in the hands of those mental health care professionals performing the roles of clinician and expert witness. In some cases, consultants also may be involved in keeping or maintaining records on firesetting youngsters as they participate in developing community-based intervention programs that serve these youngsters and their families. Advocates are less likely to be involved directly with firesetting youngsters, although they may have access to case material which may be used for convincing legislators about the importance of the problem of child firesetting. Regardless of who is keeping records on the psychosocial and firesetting histories of youngsters, it is necessary to understand what mental health care professionals can do to protect the records of these youngsters and their right to privacy.

Given the goal of protecting the privacy of firesetting youngsters by ensuring the confidentiality of their written records, the least favorable outcome that may occur is that these records are subpoenaed by a court of law. If a subpoena is received regarding a certain set of records, to comply with the law, these records must be turned over to the requesting authorities. The contents of these records then become the property of the court. In addition, if the court orders psychological evaluations, or psychological evaluations are presented on behalf of the defense, the records produced as a result of these activities are the property of the court. The best outcome is that mental health care professionals generally are responsible for what comprises these records and therefore they are in a good position to communicate in writing precisely what is necessary and accurate. Although files containing written communications regarding the behaviors of firesetting youngsters must reflect the nature of the transactions between mental health care professionals and their clients, the potential for disclosure of these records must be considered an important factor in developing and maintaining them.

Disclosure of these records is less likely to occur as a result of a subpoena or court order and more likely to occur as a result of a request by other mental health care professionals and community agencies. Because, in most instances, these are records of minors, parental permission is necessary before the information contained in these records can be disclosed. The idea of disclosing a particular portion or all of the record must be discussed with the parents of firesetting youngsters. In addition, it is advisable to obtain their permission, if they agree to disclose their record, in writing. In general, the information in these files should be disclosed when there are a specific set of circumstances and reasons suggesting that disclosure will be in the best interest of the firesetting youngsters and their families.

The final area of confidentiality that deserves attention is identification and disclosure with respect to publications and the media. Regardless of the role, clinicians, consultants, expert witnesses, and advocates are likely to face these questions as a result of writing about their professional activities, discussing their work with their colleagues, and participating, as requested, in such media activities as newspaper, television, and radio interviews. Given these activities, there are likely to be requests to disclose the identity of youthful firesetters and their families for a variety of purposes including the illustration of the problem of juvenile arson by case example, providing case material to the interested readers of newspapers and magazines, and offering in vivo testimony to government agencies regarding the participation of youngsters in the crime of arson. Although there may be a number of convincing reasons why the identity of youthful firesetters and their families should be disclosed, there are a number of reasons why careful consideration must be given before the decision is made to reveal identities.

The major responsibility of deciding to reveal the specific identity of youthful firesetters and their families rests with the parents of the families. However, it is the responsibility of mental health care professionals to inform these youngsters and their families about the potential risks of revealing their identities. Because these youngsters are likely to have been involved in firesetting incidents, the public identification of these youngsters and their families within their own community could result in negative consequences and reactions from friends, associates, and those providing services to these families.

Firesetting remains a potentially threatening behavior to many and, as such, places a stigma on those who have been identified as participating in the activity. The families, with the help of mental health care professionals, must decide whether they want to assume these risks by disclosing their identities. The potential benefit derived from disclosing identities is that other youngsters and families, suffering from problems related to firesetting, may come forth and seek the necessary help as a result of seeing or reading about families experiencing similar difficulties.

If, after careful consideration of the risks and benefits of disclosing identities for purposes of written material, interviews, or other activities where youngsters and their families will be personally identified with the problem of firesetting, the decision is made not to self-identify, then this decision must be respected. No materials regarding these cases should be disclosed. However, permission may be granted by the families for the case material to be used, if the youngsters and families are not personally identified. In these instances, a written agreement is useful between mental health care professionals and their clients stating exactly how the identity of the youngsters and their families is to be protected. In addition, if families agree to identify themselves or their youngsters with the problem of firesetting, then written permission must be obtained releasing the information. Although the parents of youthful firesetters must decide whether they want their families publicly identified with the social issue of firesetting, it is the mental health care professional's responsibility to guide them as to weighing the risks versus the potential benefits in such an important decision-making process.

Establishing and maintaining a reasonable trust regarding verbal communication, protecting the privacy of written records within legal parameters, and safeguarding youngsters' and families' identities with respect to publications and the media represent the major responsibilities of confidentiality for mental health care professionals. Depending on the specific role—clinician, consultant, expert witness, and advocate—the occasion is likely to arise in which the privacy of youthful firesetters and their families will be challenged. Aside from the issues of confidentiality between mental health care professionals and their clients, parents of youngsters often are faced with the question of whether to identify their families with the problem of youthful firesetting. Mental health care professionals must have a working knowledge of all the issues regarding confidentiality, and they must be ready to play a significant role in helping families to analyze the risks and benefits of self-identification and to place a high priority within the decision-making process on selecting the outcome that is in the best interest of youngsters and their families.

Professional Liability

The two critical aspects of professional liability that are specifically relevant to working with youthful firesetters and their families are insurance and legal considerations. Mental health care professionals assuming the roles of clinician, consultant, expert witness, and advocate must also assume the potential liabilities inherent in performing the functions associated with these roles. Those roles which provide direct services to youthful firesetters and their families are at a higher risk for incurring liability because their professional activities are related in a direct manner to the behavior of the youngsters they serve. These roles include that of clinician and expert witness. However, the roles of consultant and advocate also are at risk because the nature of their functions represent youthful firesetters in a variety of ways, such as developing and maintaining effective community-based programs or designing legislation to provide juvenile firesetters adequate treatment and protection under the law. Hence, a careful consideration of insurance and legal matters will enhance the practice of mental health care professionals who are working with youthful firesetters and their families.

Because the functions associated with the roles of clinician and expert witness usually are conducted within the private practice of mental health care professionals, it becomes important to provide some type of security for continuation of that practice. There are inherent risks in maintaining such a practice, which, in part, is comprised of working with youngsters who have histories of potentially destructive or criminal behavior. Therefore, it becomes important to consider methods for protecting the practice against possible litigation resulting from the activities of their clients. One method that must be considered by mental health care professionals is the acquisition of liability insurance. A number of companies offer such insurance, and many professional organizations, such as the American Psychological Association, offer it directly to their members. In addition to protecting the private practices of clinicians and expert witnesses, consultants often may find that the agencies with whom they contract require that they have such insurance. Currently, the annual insurance rates are reasonable with respect to adequate coverage, and payment of these rates far outweighs the potential financial risks that could be incurred as a result of litigation.

Because the professional roles of mental health care providers are expanding beyond the traditional activities of providing the direct services of evaluation and treatment to youthful firesetters and their families, there may be circumstances where the services of a legal consultant is required. Mental health care professionals must learn to recognize when it is in their best interest to seek the consultation of attorneys. For example, the role of consultant may require a signing of service contracts between mental health care professionals and the agencies for whom they will provide services. Often these contracts are "boilerplate" in nature and require little interpretation. However, for the purpose of ensuring that the role and activities of mental health care professionals are adequately represented in these contracts, it is in the best interest of consultants to seek the advice of attorneys to verify the fairness of the contracts. Mental health care professionals must be able to recognize the limits of their areas of expertise and understand the circumstances in which it is appropriate to seek the consultation of other experts, especially when it relates to protection of their legal rights.

Obtaining adequate liability insurance and understanding when it is appropriate to seek the advice of counsel are two important professional activities that will protect the integrity of the practice of mental health care providers who work with youthful firesetters and their families. It is important for mental health care professionals to be able to manage the risks associated with assuming both the traditional role of clinician and the innovative roles of consultant, expert witness, and advocate. An understanding of the potential liability and the effective management of these risks not only protects mental health care professionals, but indirectly safeguards the youngsters and families who rely on competent services to help resolve their firesetting and related problems.

Roles and Responsibilities

Table 8.2 summarizes the roles likely to be assumed by mental health providers and their relationship to the professional responsibilities incurred as a result of working with youthful firesetters and their families. The specific aspects of confidentiality and liability are indicated as to their relevance to the particular roles. Although not all of the responsibilities are associated with all of the roles, mental health care professionals are well advised to understand the

Table 8.2
The Professional Responsibilities Relevant to the Specific Roles of Mental Health Care Providers

ROLES	RESPONSIBILITIES				
	Confidentiality			Liability	
	Reasonable Trust	*Written Records*	*Publications and the Media*	*Insurance*	*Counsel*
I. Clinician	X	X	X	X	
II. Consultant		X	X	X	X
III. Expert witness	X	X	X	X	X
IV. Advocate		X	X	X	X

entire spectrum of potential responsibilities. The ability to provide effective services to youthful firesetters and their families depends on the successful execution by mental health care providers of these multiple roles and related professional responsibilities.

WORKING TOWARD A COMMON SET OF GOALS

The multiple roles, related functions, and responsibilities assumed by mental health care professionals suggest that there are at least four major types of goals to be considered for advancing knowledge in the field of child firesetting. These four goals include evaluation, intervention, prevention, and advocacy. Each of these four goals is comprised of specific objectives which represent particular professional activities. Participation in these professional activities is likely to reduce the incidence of fireplay, firesetting, and juvenile-related arson and increase the availability of an effective service delivery system to help youthful firesetters and their families.

The goal of evaluation refers to mental health care professionals providing psychosocial evaluations with a special expertise in conducting a comprehensive history of fire behavior. In addition, mental health care professionals must be able to analyze and interpret their findings so that accurate linkages can be inferred between the observed fire behavior and the psychosocial determinants. Moreover, in the special-circumstance evaluations, such as those to be utilized either by the prosecution or by the defense in a court of law, clinicians must be able to identify the characteristics of mental responsibility and relate them to the criminal behavior of arson. Conducting, analyzing, interpreting, and presenting data comprise the major professional activities of delivering the services of effective evaluation.

The goal of intervention can be accomplished by mental health care providers participating in at least two major professional activities. First, clinicians can provide the direct services of psychotherapy to youthful firesetters and their families. Second, mental health care professionals can act as consultants and offer their skills in helping to develop community-based intervention programs. In the case of either the direct or indirect delivery of effective intervention, clinicians and consultants can make a significant contribution to eliminating the problem of firesetting as well as remediating the accompanying psychosocial determinants.

The goals of evaluation and intervention are important once the firesetting problem emerges. However, the goal of prevention is the unique method of ensuring that youngsters' naturally occurring interest in fire will be expressed in terms of fire-safe, as opposed to fire-risk, behaviors. Mental health care professionals, assuming roles as educators and program development experts, can promote and implement age-appropriate fire education and safety programs within schools, fire departments, and other relevant community agencies. Participation in prevention efforts is likely to reduce the incidence of nonproductive firestarting and juvenile-related arson.

The goal of advocacy involves a variety of innovative professional activities for mental health care providers. First, the role of educator can be assumed not only to train other professionals to work on the problem of child firesetting, but to educate the general public, through activities such as conducting seminars or working with the media, on what members of the community can do to prevent juvenile-related arson. Second, mental health care professionals can participate in local, state, and federal legislative processes to ensure that youngsters accused of arson are treated humanely and given adequate and fair protection under the law. Finally, mental health care professionals can initiate and support research projects that aim at understanding not only why youngsters become involved in firesetting, but what intervention strategies are most effective for eliminating nonproductive firestarting and remediating the accompanying psychopathology. The goal of advocacy and the related professional activities will encourage a "critical mass" of mental health care providers to break new ground in understanding the problem of child firesetting.

Table 8.3 summarizes the common set of goals and associated objectives toward which mental health care professionals must work to reduce the incidence of fireplay, firesetting, and juvenile-related arson. Evaluation, intervention, prevention, and advocacy represent professional activities that will advance the "edge of the field" with respect to eliminating youthful firesetting behavior and remediating

Table 8.3
A Common Set of Goals and Objectives

Goals		Objectives
I. Evaluation	A.	Psychosocial evaluation with a special expertise in assessing fire behavior history
	B.	Court evaluation to identify mental responsibility
II. Intervention	A.	Direct service delivery of psychotherapy
	B.	Developing community-based intervention programs
III. Prevention	A.	Promote and implement age-appropriate fire education and safety programs
IV. Advocacy	A.	Public education activities focused on prevention
	B.	Participation in local, state, and federal legislative processes to protect the rights of firesetting youngsters
	C.	Support and conduct research directed at why youngsters firestart and the most effective intervention strategies to eliminate the behavior and remediate the accompanying psychopathology

the accompanying psychosocial determinants. It is through the dedicated work of achieving these goals, and assuming the related roles and responsibilities, that dramatic progress can be accomplished in resolving the problem of child firesetting.

SUMMARY

Mental health care professionals can assume multiple roles, functions, and responsibilities in working with youthful firesetters and their families. There are at least four major roles—clinician, consultant, expert witness, and advocate—which carry with them specific functions to be performed by mental health care professionals. The role of clinician is characterized by the direct service functions of psychological evaluation and treatment. Two types of consultant activities are developing community-based intervention programs and providing education and training services to other professionals interested in learning how to work with youthful firesetters and their families. The specific activities of an expert witness focus on the skills of applying psychosocial data to identify mental responsibility as it relates to the crime of arson. The role of advocate consists of activities that can result in changes in social policy regarding youthful firesetting, such as direct participation in political and governmental processes and execution of applied research projects. In addition to these roles and functions, there are two major categories of professional responsibilities—confidentiality and liability—which mental health care providers must consider when working on behalf of youthful firesetters and their families. The three critical issues concerning confidentiality are the notion of reasonable trust regarding verbal communication, the privacy of written records, and identification and disclosure considerations regarding publications and the media. The two essential aspects of professional liability for mental health care providers are obtaining adequate insurance to cover the value of their practice and understanding when it is appropriate to seek the advice of counsel. These multiple roles, functions, and responsibilities suggest that there is a common set of goals—evaluation, intervention, prevention, and advocacy—representing the primary professional activities in which mental health care providers must engage to advance knowledge in understanding the psychology of child firesetting.

References

Abrams, G. (1985). Treating kids who play with fire. *Los Angeles Times*, April, 1-3.

Akiyama, Y., & Pfeiffer, P. C. (1984). Arson: A statistical profile. *FBI Law Enforcement Bulletin, 53*, 8–14.

American Psychiatric Association (1980). *Diagnostic and statistical manual of mental disorders (3rd ed.)*. Washington, DC: American Psychiatric Association.

Awad, G. A., & Harrison, S. I. (1976). A female firesetter: A case report. *The Journal of Nervous and Mental Disease, 163*, 432–437

Bender, L. (1959). Children and adolescents who have killed. *American Journal of Psychiatry, 116*, 510–513.

Benians, R. C. (1981). Conspicuous firesetting in children. *British Journal of Psychiatry, 139*, 366.

Birchill, L. E. (1984). Portland's firesetter program involves both child and family. *American Fire Journal, 23*, 15–16.

Block, J. H., & Block, J. (1975). *Fire and children: A study of attitudes, behaviors and maternal teaching strategies*. Technical Report for Pacific Southwest Forest and Range Services, United States Department of Agriculture.

Block, J. H., Block, J., & Folkman, W. S. (1976). *Fire and children: Learning survival skills*. USDA Forest Service Research Paper PSW-119.

Block, J. H., Block, J., & Folkman, W. S. (1976). *Fire and children. Learning survival skills*. Pacific Southwest Forest and Range Experiment Station, USDA Forest Service Research Paper, PSW-119.

Broling, L., & Brotman, C. (1975). A fire-setting epidemic in a state mental health care center. *American Journal of Psychiatry, 132*, 946–950.

Bumpass, E. R., Fagelman, F. D., & Brix, R. J. (1983). Intervention with children who set fires. *American Journal of Psychotherapy, 37*, 328–345.

Buros, O. K. (1972). *Seventh mental measurements yearbook*. Rutgers, NJ: Rutgers University Press.
California Division of Forestry (1977). *Children caused fires*. Riverside Fire Prevention Unit.
California Penal Code (1971). Section 676.
California Penal Code (1979). Section 451.
California Penal Code (1981). Section 26.
California State Code (1986). Chapter 1529, Juvenile arson and firesetting.
Carstens, C. (1982). Application of a work penalty threat in the treatment of a case of juvenile fire setting. *Journal of Behavior Therapy and Experimental Psychiatry, 13,* 159–161.
The Children's Television Workshop (1982). *Sesame Street fire safety resource book*. New York: The Children's Television Workshop.
Cole, R. E., Laurentis, L. R., McAndrews, M. M., McKeever, J. M., & Schwartzman, P. (1984). *Juvenile firesetter intervention. A report of the Rochester, New York, Fire Department Fire Related Youth Program Development Project*. Rochester, NY: The New York State Department of State Office of Fire Prevention and Control.
Committee on the Judiciary (1985). *The problem of arsons committed by juveniles*. Washington, DC: U.S. Government Printing Office.
Eisler, R. M. (1974). Crisis intervention in the family of a firesetter. *Psychotherapy: Research, Theory and Practice, 9,* 76–79.
Federal Bureau of Investigation (1977). *Uniform crime reports*. Washington, DC: U.S. Government Printing Office.
Federal Bureau of Investigation (1982). *Uniform crime reports*. Washington, DC: U.S. Government Printing Office.
Federal Bureau of Investigation (1983). *Uniform crime reports*. Washington, DC: U.S. Government Printing Office.
Federal Bureau of Investigation (1985). *Uniform crime reports*. Washington, DC: U.S. Government Printing Office.
Federal Emergency Management Agency (1979). *Interviewing and counseling juvenile firesetters. The child under seven years of age*. Washington, DC: U.S. Government Printing Office.
Federal Emergency Management Agency (1983). *Juvenile firesetter handbook: Dealing with children ages 7–14*. Washington, DC: U.S. Government Printing Office.
Fine, S., & Louie, D. (1979). Juvenile firesetters: Do agencies help? *American Journal of Psychiatry, 136,* 433–435.
Fineman, K. R. (1980). Firesetting in childhood and adolescence. *Psychiatric Clinics of North America, 3,* 483–500.
Folkman, W. S. (1966). *Children with matches. Fires in the Angeles National Forest Area*. USDA Forest Service Research Note, PSW-109.
Freud, S. (1932). The acquisition of power over fire. *International Journal of Psychoanalysis, 13,* 405–410.
Gaynor, J. (1985). Child and adolescent fire setting: Detection and intervention. *Feelings and Their Medical Significance, 27,* 1–10.
Gaynor, J., Huff, T. G., & Karchmer, C. L. (1986). The linkages between childhood firestarting and adult arson crime: A secondary analysis of convicted arsonists' retrospective reports. San Francisco: National Firehawk Foundation, Research Report No. 1.

Gaynor, J., McLaughlin, P. M., & Hatcher, C. (1983). *The Firehawk children's program. A working manual.* San Francisco: National Firehawk Foundation.

Gladston, R. (1972). The burning and the healing of children. *Psychiatry, 35,* 57–66.

Gold, L. (1962). Psychiatric profile of a firesetter. *Journal of Forensic Sciences, 7,* 404–417.

Gruber, A. R., Heck, E. T., & Mintzer, E. (1981). Children who set fires: Some background and behavioral characteristics. *American Journal of Orthopsychiatry, 51,* 484–488.

The Hartford Insurance Group (undated). *Prevent fire before it starts.* (3rd Revision) 81932.

Heath, G. A., Gayton, W. F., & Hardesty, V. A. (1976). Childhood firesetting. *Canadian Psychiatric Association Journal, 21,* 229–237.

Heath, G. A., Hardesty, V. A., Goldfine, P. E., & Walker, A. M. (1983). Childhood firesetting: An empirical study. *Journal of the American Academy of Child Psychiatry, 22,* 370–374.

Holland, C. J. (1969). Elimination by the parents of fire-setting behavior in a 7-year-old boy. *Behavior Research and Therapy, 7,* 135–137.

Hurley, W., & Monahan, T. (1969). Arson: The criminal and the crime. *British Journal of Criminology, 9,* 4–21.

Jones, F. D. E. (1981). Therapy for firesetters. *American Journal of Psychiatry, 138,* 261–262.

Jones, R. T., Kazdin, A. E., & Haney, J. I. (1981). Social validation and training of emergency fire safety skills for potential injury prevention and life-saving. *Journal of Applied Behavior Analysis, 14,* 249–260.

Kafry, D. (1978). *Fire survival skills: Who plays with matches?* Technical Report for Pacific Southwest Forest and Range Service, U.S. Department of Agriculture.

Kafry, D., Block, J. H., & Block, J. (1981). *Children's survival skills: A basis for functioning in society.* Final Report Prepared for The Maternal and Child Health and Crippled Children's Services Research Program. Bureau of Community Health Services, Rockville, Maryland.

Kaufman, I., Heins, L., & Reiser, D. (1961). A re-evaluation of the psychodynamics of firesetting. *American Journal of Orthopsychiatry, 31,* 123–136.

Kolko, D. J. (1983). Multicomponent parental treatment of firesetting in a developmentally disabled boy. *Journal of Behavior Therapy and Experimental Psychiatry, 14,* 349–353.

Kolko, D. J., Kazdin, A. E., & Meyer, E. C. (1985). Aggression and psychopathology in childhood firesetters: A controlled study of parent and child reports. *Journal of Consulting and Clinical Psychology, 53,* 377–385.

Koret, S. (1973). Family therapy as a therapeutic technique in residential treatment. *Child Welfare, 52,* 235–246.

Kuhnley, E. J., Hendren, R. L., & Quinlan, D. M. (1982). Firesetting by children. *Journal of the American Academy of Psychiatry, 21,* 560–563.

Lewis, N. O. C., & Yarnell, H. (1951). Pathological firesetting (pyromania). *Nervous and mental disease,* Monograph (No. 82). Nicholasville, KY: Collidge Foundation.

Macht, L. B., & Mack, J. E. (1968). The firesetter syndrome. *Psychiatry, 31,* 277–288.

Madanes, C. (1981). *Strategic family therapy.* San Francisco: Jossey-Bass.

McGrath, P., Marshall, P. G., & Prior, K. (1979). A comprehensive treatment program for a fire setting hospital. *Journal of Behavior Therapy and Experimental Psychiatry, 10,* 69–72.

McKerracher, D. W., & Dacre, J. I. (1966). A study of arsonists in a special security hospital. *British Journal of Psychiatry, 112,* 1151–1154.
McKinney, C. D. (1983). *Houston juvenile firesetters prevention program.* Houston, TX: Unpublished Report.
McLaughlin, P. (1983). Statistics on the San Francisco Fire Department's juvenile firesetters program. Personal communication.
Minuchin, S. (1974). *Families and family therapy.* Cambridge, MA: Harvard University Press.
Moos, R., & Insel, P. M. (1974). *Issues in social ecology.* Palo Alto, CA: National Press Books.
Moscrip, C. G. (1986). Personal communication regarding legal consultation on youngsters convicted of arson.
National Committee on Property Insurance (1984). *Juvenile firesetters programs.* Boston: National Committee on Property Insurance.
The National Fire Protection Agency (1979). *Learn Not to Burn Curriculum.* Boston: The National Fire Protection Agency.
Nielsen, J. (1970). Criminality among patients with Klinefelter's syndrome and the XYY syndrome. *British Journal of Psychiatry, 117, 365.*369.
Patterson, G. R. (1976). The aggressive child: Victim and architect of a coercive system. In L. A. Hamerlynck, L. C. Handy, & E. J. Mash (Eds.), *Behavior modification and families* (pp. 267–316). New York: Brunner/Mazel.
Patterson, G. R. (1978). The aggressive child. In G. R. Patterson & J. B. Reid (Eds.), *Systematic common sense.* Eugene, OR: Castalia Press.
Patterson, G. R., Reid, J. B., Jones, R. R., & Corger, R. E. (1975). *A social learning approach to family intervention. Vol. I: Families with aggressive children.* Eugene, OR: Castalia Publishing Co.
Pollack, S. (1980). Arson prosecution. Sacramento, CA: California District Attorneys Association.
Rider, A. (1980). The firesetter. A psychological profile. Washington, DC: Federal Bureau of Investigation.
Ritvo, E., Shanok, S. S., & Lewis, D. O. (1982). Firesetting and nonfiresetting delinquents: A comparison of neuropsychiatric, psychoeducational, experiential and behavioral characteristics. *Child Psychiatry and Human Development, 13,* 259–267.
Rodrigue, G. (1982). Kid torches: Babes that burn. A successful method of counseling juvenile firesetters in Dallas. *Firehouse,* 49–52.
Sakheim, G. A., Vigdor, M. G., Gordon, M., & Helprin, L. M. (1985). A psychological profile of juvenile firesetters in residential treatment. *Child Welfare, 64,* 453–476.
Schiller, N. G., & Jacobson, M. (1984). *Stress-induced arson: An example of stress-induced crime.* New York: The City of New York Arson Strike Force.
Siegel, L. (1957). Case study of a thirteen-year-old firesetter. A catalyst in the growing pains of a residential inpatient unit. *American Journal of Orthopsychiatry, 27,* 396–410.
Siegelman, E. (1969). *Children who set fires. An exploratory study.* Conducted for the Resources Agency of California, Department of Conservation, Division of Forestry.

Siegelman, E. Y., & Folkman, W. S. (1971). *Youthful firesetters: An exploratory study in personality and background*. Springfield, VA: USDA Forest Service.

Stawar, T. L. (1976). Fable mod: Operantly structured fantasies as an adjunct in the modification of fire-setting behavior. *Journal of Behavior Therapy and Experimental Psychiatry, 7*, 285–287.

Stekel, W. (1924). *Peculiarities of behavior (Vol. II)*. New York: Boni and Liveright.

Stewart, M. A., & Culver, K. W. (1982). Children who set fires. The clinical picture and follow-up. *British Journal of Psychiatry, 140*, 357–363.

Strachan, J. C. (1981). Conspicious firesetting in children. *British Journal of Psychiatry, 138*, 26–29.

Teague, P. (1978). Action against arson. *Fire Journal*, 46.

US Department of Commerce, National Fire Prevention and Control Administration, National Fire Data Center (1978). *Highlights of fire in the United States: Deaths, injuries, dollar loss, and incidents at the national, state and local levels*.

Vandersall, J. A., & Weiner, J. M. (1970). Children who set fires. *Archives of General Psychiatry, 22*, 63–71.

Vreeland, R. G., & Waller, M. B. (1979). *The psychology of firesetting: A review and appraisal*. Washington, DC: U.S. Department of Commerce, National Bureau of Standards.

Welsh, R. S. (1971). The use of stimulus satiation in the elimination of juvenile firesetting behavior. In A. M. Graziano (Ed.), *Behavior therapy with children*. Chicago: Aldine Publishers.

Whittington, C., & Wilson, J. R. (1980). Fat fires: A domestic hazard. In D. Canter (Ed.), *Fires and human behavior* (pp. 97–115). Chichester, England: John Wiley and Sons.

Wilmore, D. W., & Pruitt, B. A. (1972). Toward taking the fat out of fire. *Medical World News, 27*, 16.

Wolford, M. R. (1972). Some attitudinal, psychological and sociological characteristics of incarcerated arsonists. *Fire and Arson Investigator*, 22(4), 1–30; 22(5), 1–26.

Wooden, W., & Berkey, M. L. (1984). *Children and arson. America's middle class nightmare*. New York: Plenum.

Yarnell, H. (1940). Firesetting in children. *American Journal of Orthopsychiatry, 10*, 272–286.

Author/Source Index

Subject Index